T0183508

NEXUS NETWORK JOURNAL   Architecture and Mathematics

Nexus Network Journal (NNJ) is a peer-reviewed journal for researchers, professionals and students engaged in the study of the application of mathematical principles to architectural design. Its goal is to present the broadest possible consideration of all aspects of the relationships between architecture and mathematics, including landscape architecture and

E-mail: kwb@kimwilliamsbooks.com

E-mail: syld@kimwilliamsbooks.com

E-mail: alessandra.capanna@uniroma1.it

E-mail: tgsalgado@perspectivegeometry.com

The Chinese University of Hong Kong

School of Architecture and Built Environment
Faculty of Engineering and Built Environment

E-mail: michael.ostwald@newcastle.edu.au

Vera Spinadel
The Mathematics & Design Association
José M. Paz 1131 - Florida (1602), Buenos Aires, Argentina
E-mail: vspinade@fibertel.com.ar

Igor Verner
The Department of Education in Technology and Science
Technion - Israel Institute of Technology
Haifa 32000, Israel
E-mail: ttrigor@techunix.technion.ac.il

Stephen R. Wassell
Department of Mathematical Sciences
Sweet Briar College, Sweet Briar, Virginia 24595, USA
E-mail: wassell@sbc.edu

João Pedro Xavier
Faculdade de Arquitectura da Universidade do Porto
Rua do Gólgota 215, 4150-755 Porto, Portugal
E-mail: jpx@arq.up.pt

Instructions for Authors

Authorship

Submission of a manuscript implies:
- that the work described has not been published before;
- that it is not under consideration for publication elsewhere;
- that its publication has been approved by all coauthors, if any, as well as by the responsible authorities at the institute where the work has been carried out;
- that, if and when the manuscript is accepted for publication, the authors agree to automatically transfer the copyright to the publisher; and
- that the manuscript will not be published elsewhere in any language without the consent of the copyright holder.

Exceptions of the above have to be discussed before the manuscript is processed. The manuscript should be written in English.

Submission of the Manuscript

Material should be sent to Kim Williams
via e-mail to: kwb@kimwilliamsbooks.com
or via regular mail to: Kim Williams Books,
Corso Regina Margherita, 72,
10153 Turin (Torino), Italy

Please include a cover sheet with name of author(s), title or profession (if applicable), physical address, e-mail address, abstract, and key word list.

Contributions will be accepted for consideration to the following sections in the journal: research articles, didactics, viewpoints, book reviews, conference and exhibits reports.

Final PDF files

Authors receive a pdf file of their contribution in its final form. Orders for additional printed reprints must be placed with the Publisher when returning the corrected proofs. Delayed reprint orders are treated as special orders, for which charges are appreciably higher. Reprints are not to be sold.

Articles will be freely accessible on our online platform SpringerLink two years after the year of publication.

Nexus Network Journal

ARCHITECTURE, MATHEMATICS AND PERSPECTIVE

Tomás García-Salgado, Guest Editor

VOLUME 12, NUMBER 1
Spring 2010

KIM WILLIAMS BOOKS

Nexus Network Journal
Vol. 12
No. 1
Pp. 1-162
ISSN 1590-5896

CONTENTS

The seminal idea for this special issue of the *Nexus Network Journal* dedicated to Perspective was born while drinking coffee with Kim Williams during a recess of the Nexus 2006 conference in Genoa. Two years later, at Nexus 2008 in San Diego, Kim and I released the Call for Papers with the goal of bringing together original investigations regarding perspective, looking for new insights that might enrich the knowledge of this science. Now, it is the reader's turn to judge if we have achieved our goal. To begin, I would like to introduce each one of the contributors.

My curiosity was piqued when Volker Hoffmann told me by e-mail: "working on my contribution for your edition of *NNJ*, I made a nice discovery which changes the common history of perspective." Once I received his article on "Giotto and Renaissance Perspective", I became aware of the importance that his discovery could have. In my opinion, his insightful writing is going to raise many questions among the scholars of perspective, or at least it is going to make them reconsider when and where on the timeline of the history of art the first rigorous application of perspective will have to be placed. Here, Hoffmann provides the answer to this question by means of his geometrical analysis of the left *coretto* painted by Giotto in the Arena Chapel (Padua), a remarkable analysis that could bridge the gap between Gothic and Renaissance painting.

In his article, "Perspective, a Visionary Process: The Main Generative Road for Crossing Dimensions", Celestino Soddu brings together perspective with other geometries to explore architectural design. He starts by exploring space from a fixed point, and then moving all around the observer so that he is able to render endless points of view of one target. With the help of the computer program that he himself developed, Celestino has achieved the perspective practitioner's dream of, that is, the total visualization of the object. This will be clear to the reader from the illustration of his project to create a 360° spherical perspective of the Tower of Babel. In Celestino's own words: "We could see, together in the same drawing, the front and the rear, the right, the left, above and below." His new approach in design, called Generative Art, states that the idea is more important than the product, a principle that is well portrayed in the conclusion of this interesting article.

In her article "Perspective in a Box", Agnes Verweij analyzes how a Dutch peep-show-box works geometrically. Most Dutch perspective boxes were made during the seventeenth century but only a few have survived until today. The extant boxes are on exhibit at the National Museum of Denmark (Copenhagen), the Museum Bredius (The Hague), and The National Gallery of London. Included in the Bredius collection is a triangular box, leading the spectator to wonder how it is possible to place the scene of a rectangular space within a triangular shape, while in the London collection, the internationally known box by Hoogstraten shows an atypical disposition of two peepholes through which the opposing views of the interior of a Dutch house can be observed. Dutch artists used real architectural scenes to captivate people's imaginations; thus, when looking through the box's peephole the observer was able to recognize to which building the scene belongs. The cut-out of the baker's box, discussed at the end of Agnes's article, shows us that Danish people still carry on this unique tradition today.

The question of whether Ezekiel's vision of the temple refers to Solomon's temple or an ideal, unbuilt temple is a controversial one, although most scholars agree with the latter theory. An explicative note in the Ryrie Study Bible points out: "The description is not of

DOI 10.1007/s00004-010-0025-5; *published online* 10 March 2010
© 2011 Kim Williams Books, Turin

Solomon's Temple, the specifications being different and larger" (Moody Press, Chicago, 1994, p. 1282). In her article, "Juan Bautista Villalpando and the Nature and Science of Architectural Drawing", Tessa Morrison addresses this controversial issue, and states clearly that: "Villalpando claimed that Ezekiel's vision was the Temple of Solomon and he made no distinction between Solomon's Temple and the vision of Ezekiel." Here, the aim of studying Villalpando's treatise, *In Ezechielem Explanationes*, is not the controversy itself; rather, it is the analysis of the ideas about architectural drawing contained in Villalpando's treatise. Tessa's proficiency in Latin gives us an opportunity of reading a first-hand interpretation of Villalpando's exegesis on the book of Ezekiel.

In the article, "Perspective versus Stereotomy: From Quattrocento Polyhedral Rings to Sixteenth-Century Spanish Torus Vaults", José Calvo-López and Miguel Ángel Alonso-Rodríguez explain the principles of orthographic projections and perspective, and what features they have in common. Both Piero della Francesca's treatise *De prospettiva pingendi* and Gil Rodríguez de Junterón's design for the Recapilla chapel in the Cathedral of Murcia will show the reader that stereotomy and perspective do indeed share common principles. While the stonemasons needed orthographic projections to determine the voussoirs of an arch at full scale, Piero used these projections at a random scale to construct a perspective. The interplay of perspective and stereotomy is made clear in the figure of Piero's construction for a rotated cube applied to a *mazzocchio*. Whether Spanish architects may have been influenced by the Italian Quattrocento or by the Gothic tradition, is another question at the core of this article.

In my own article, "The Sunlight Effect of the Kukulcán Pyramid, or the History of a *Line*," I explain the geometrical features of the pyramid that produce its unique effects of light and shadow. I started by formulating a hypothesis about how the base of the pyramid was laid out, and this in turn spurred me to propose a method for laying out a square (of a given side) without using Euclidean right angles. I present an experiment that was actually performed on the UNAM campus to prove the feasibility of what I call the Mayan method for laying out a square. As we know, many scholars have repeatedly referred to the singular orientation of the diagonal line of the pyramid base but without fully understanding its meaning. The meaning of this line, which I am convinced is of paramount importance, is that of being the "prime line" from which the pyramid was built. Despite the fact that the field of Mayan studies is quite extensive, little is known about the geometry of their buildings in regard to the methods by which the builders achieved their forms. These and many other questions, such as whether their observatories were built at day or at night, will have to be investigated in order to achieve a better understanding of Mayan geometry.

John Hatch's article, "Some Adaptations of Relativity in the 1920s and the Birth of Abstract Architecture," bridges the gap between the traditional concept of space and the concept of space-time as a higher dimension. The Renaissance painters conquered the third dimension; the breakthrough of the Theory of Relativity brought a new dimension, and with it, the vision of new geometries for the arts. The early twentieth-century artists eagerly embraced the works of Minkowski, Lorentz, Poincaré, and Einstein, pursuing how to incorporate time and space in painting and architecture. Hatch brings to the core of his article the artworks of El Lissitzky and Theo Van Doesburg. The effort to conquer time and space is well portrayed in these words: "However, a unique feature of Van Doesburg's design is that there is never one fixed point from which one can define the whole of the structure. Every vantage point provides a unique view that is never repeated twice throughout. In other words, there is no defining moment, no fixed or absolute point, and thus Van Doesburg achieves an inventive type of completely relativistic, Dadaist type of architecture. It embodies a notion we will encounter with Lissitzky, that every point in space is related to a unique moment in time." In my opinion, the "relativistic" vision of space does not neglect the traditional vision of perspective because it comes from it. Lissitzky's Suprematist works, inspired by Minkowski's space-time diagram, describe the struggle of the oblique presentation of the x- and y-axes against the orthogonal ones in a 2D composition, in manner that is similar to what Van Doesburg did for the ceiling of the University Hall (Amsterdam, 1923); and so on until reaching four dimensions.

This special issue also includes book reviews related to perspective. Samuel Edgerton comments on his own book, *The Mirror, the Window, and the Telescope: How Renaissance Linear Perspective Changed Our Vision of the Universe*. João Pedro Xavier reviews Kirsti Andersen's book *The Geometry of an Art. The History of Perspective from Alberti to Monge*. I myself review *Forma y Representación. Un Análisis Geométrico* by Javier Navarro de Zuvillaga.

To conclude this letter, I would like to express my gratitude to Kim Williams, and also to each one of the authors, for their efforts in accomplishing the present Special Issue of the *Nexus Network Journal* dedicated to Perspective.

Volker Hoffmann

Sterntalerweg 4, D-97084
Würzburg, Germany
volker.hoffmann@ikg.unibe.ch

Keywords: Giotto,
Brunelleschi, Masaccio,
Perspective, costruzione
legittima

Research

Giotto and Renaissance Perspective

Abstract. A careful geometric examination of the blind arcades (*coretti*) depicted in Giotto's fresco on the choir wall in the Arena Chapel in Padua shows that they were designed and painted according to the rules of what I term "progressive *costruzione legittima*" and thus represent simulations of visual images. Because no images of this type have come down from Classical Antiquity and because the literary references remain silent in this respect, the *coretti* must be considered, according to today's knowledge, the oldest monuments manifesting the application of the *costruzione legittima*. This means the history of the central (linear) perspective must be rewritten. In any case it was not a Renaissance invention. I expressly agree with the researchers who see Giotto's painting in conjunction with the findings of the Scholastic "optics specialists" (such as Grosseteste, Witelo, Bacon), who all stood with their feet firmly planted on the ground of Euclid's rigidly geometrically conceived visual theory and its Arab commentators.

Introduction

My work on the history of perspective has already been summarized in an essay on Brunelleschi [Hoffmann 1990-1992] as well as one on Masaccio [Hoffmann 1996]: They form the basis for this article on Giotto (ca. 1266-1337). In those earlier essays I made a critical study of the academic literature and do not need to repeat that here. My article on Giotto will provide findings that cannot be gained from the literature, so that I will restrict myself to just a few bibliographical references. Apart from that I refer to the well-known bibliographies on the topic of perspective [Vagnetti 1979; Veltman 1986; Alberti 2000; Sinisgalli 2006].

If one defines "perspective" as a method of representing a three-dimensional object on a plane, one comes to the conclusion, in agreement with known works in art history, that there are many perspectives: parallel and central perspective, axonometry, vanishing point perspective, reverse perspective and many others, above all, however, the mixed forms. Here we are dealing with the (monocular) central perspective, also called linear perspective, which can be geometrically constructed. This central perspective – this is lexicon knowledge – was invented in Florence in the early fifteenth century. The key works are 1) Brunelleschi's (1377-1446) *vedute* of the Florentine Baptistery and the Piazza della Signoria. These panels have not survived, but we know of them from the report by Antonio Manetti, Brunelleschi's first biographer, in which he writes, inter alia, that Brunelleschi painted them in his youth, which perhaps makes reference to the years before rather than after 1400; 2) Masaccio's (1401-1428) *Trinity* fresco in S. Maria Novella in Florence (after 1426) (fig. 1). From the point of perspective the work is divided into two: the tomb of Adam below and the donor couple kneeling on a platform above it lie in front of the wall; behind the wall a barrel vault canopy opens up above, where God the Father appears with the figures of the Holy Trinity. (It is likely that Brunelleschi was consulted in the conception of this work.)

Nexus Network Journal 12 (2010) 5–32   NEXUS NETWORK JOURNAL – VOL.12, No. 1, 2010 **5**
DOI 10.1007/s00004-010-0015-7; *published online* 9 February 2010
© 2011 Kim Williams Books, Turin

Fig. 1. Masaccio, Trinity fresco (Florence, S. Maria Novella)

What is the difference between these two prototypes of central perspective painting? Brunelleschi represents real bodies on the plane; Masaccio simulates bodies on the plane, as if they were real. This difference is fundamental, yet is scarcely mentioned in the literature on perspective. It is important now to ask how Brunelleschi produced his *vedute*. I have answered this question with the comment that this was only possible with a perspective apparatus and have decided in favor of the apparatus that Alberti first described (but incompletely) in 1435 as the "velo" (a grid of threads), and Albrecht Dürer first illustrated in a print in 1538 (fig. 2).

Fig. 2. Albrecht Dürer, *Der Zeichner des liegenden Weibes*, woodcut 1538

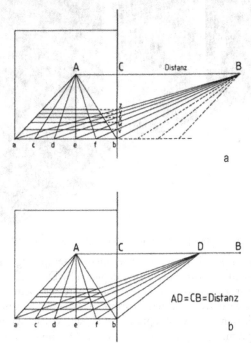

Fig. 3. The *costruzione legittima*: a, above) visual ray construction; b, below) distance point construction

And how did Masaccio construct the perspective of the *Trinity*? My answer: according to the rules of the *costruzione legittima*. This is not a fifteenth-century term, but rather appears for the first time in the seventeenth century (according to Bätschmann [Alberti 2000: 127, note 192]), despite the fact that up to today it still has no clear and generally recognized definition. Fig. 3 displays the two commonly represented variants of the *costruzione legittima*: a) the so-called "visual ray method", which Panofsky [1915] reconstructed graphically from Alberti's description[1]; and b) its simplification and transformation into the "distance point construction".

Both methods solve the task of perspectively deforming a square, represented by the line segment ab in fig. 3, as it would be seen in reality from the eye level and eye distance of a specified observer. The relationship between these two methods were first described exactly by Vignola [1583],[2] who called them *prima regola* and *seconda regola* and determined that the same result could be achieved with both. Therefore, it is nonsensical to declare one method *costruzione legittima* but not the other.[3] No, both constructions are *legittime* and deserve this attribute because they represent geometric methods that simulate the natural (monocular) visual image.

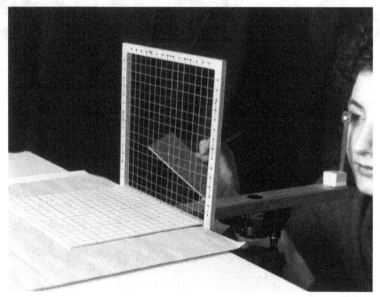

Fig. 4. The perspective apparatu: fixing the visual image

This can be demonstrated with an experimental setup (fig. 4). A perspective apparatus of the kind used by Brunelleschi, and depicted by Dürer (fig. 2), consisting of a frame with a grid of threads (*velo*) and an eyepoint, here represented by a small stand perforated with a hole securely mounted in front of it, concretely illustrate important theorems of the Euclidean theory of optics: visual ray, visual pyramid, cross-section through the visual pyramid, eyepoint and distance. The observer looks through the eyepoint (= apex of the visual pyramid) and has the task of drawing the outline of the square board lying behind the *velo* (= cross-section through the visual pyramid) onto a little panel with the same, but graphically drawn grid. Through the *velo* he sees the real square board as a trapezoid and can transfer this visual image point by point to his little panel by using the threads as reference lines: This is fixing of the visual image based on

Euclid's geometric *Optics*. As I demonstrated elsewhere [Hoffmann 1990-1992], and as far as we can know at present, Brunelleschi was the first to have the idea of intersecting the Euclidean visual pyramid, with its base on the seen object and its apex in the eye of the observer, by using the grid of threads and fixing the visual image with the threads (reference lines). In addition, the perspective apparatus can serve as a measuring device [Hoffmann 2002]. For example, if I know the width of the object seen, I can determine its height and distance. The perspective apparatus does in principle the same thing that today's photography and photogrammetry now do much, much better, namely they fix and measure the monocular visual image.

One can now replace the "visual ray" in the experiment that falls from the eyepoint onto the top edge of the square board by a thread, trace it through the *velo* and transfer this position by hand to the little square panel; the trapezoid will then have the same outline as described above. There is an even easier method (fig. 5).

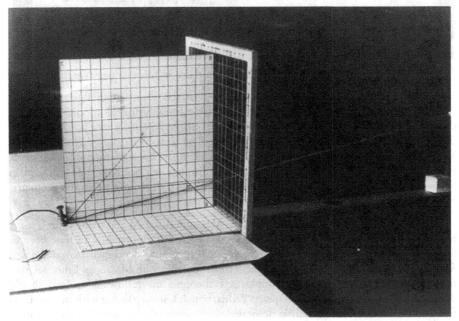

Fig. 5. The perspective apparatus: simulating the visual image ("visual ray construction")

If one places the little panel at a right angle to the grid of threads, marks a point on it on which the thread (visual ray) penetrates the grid, and draws a horizontal line through this point, then the top edge of the square board is represented in perspective. A "visual image" has been created without using the eye. The experimental arrangement corresponds exactly to fig. 3a (and hence Vignola's *regola prima*), whereby the perpendicular line through C represents the grid of threads. This geometric construction in which the genetic relationship to the perspective apparatus remains recognizable enables one to simulate a visual image; fig. 6 and fig. 3b show the method simplified into the "distance point construction" (Vignola's *regola seconda*). The grid of threads is turned around its central axis by 45° and replaced by the little square panel. The (thread) visual ray intersects the triangle ABF directly at the level of the apex of the trapezoid. (We will disregard the geometrical markings for the moment.)

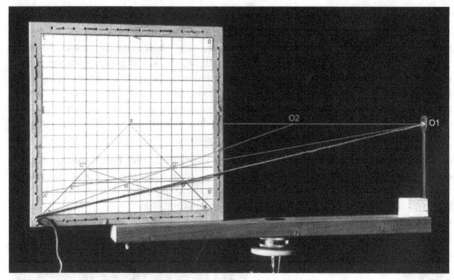

Fig. 6. The perspective apparatus: simulating the visual image ("distance point construction")

To summarize, the perspective constructions in accordance with figs. 3a and 3b (Vignola's *regola prima e seconda*) are of particular cultural historical importance in that they enable one to simulate the monocular visual image according to Euclidean optics; they are thus also the basis for illusionist painting. The term *costruzione legittima* should be reserved for this type of painting (and for it alone). This construction is *legittima*, however, only then when the task is to deform a square perspectively. We will explain below how a rectangle can also be deformed perspectively according to optic-geometric rules, despite these universal laws.

In my article in 1996 I demonstrated that Masaccio designed and painted his *Trinity* based on the *costruzione legittima*. I envisaged the historical procedure to have been as follows: First Brunelleschi invented the perspective apparatus, developed the *costruzione legittima* from it and then constructed the *Trinity* together with Masaccio. I now see that this was not at all as clear and simple as I thought; the following analysis should demonstrate this. In the Arena Chapel at Padua, which Giotto painted with his assistants from 1303-1305, the choir wall is decorated with perspectively painted architecture in the form of a "Syrian arch" above four pilasters. In the lower zone theatre-box-like blind arcades have been placed on the left and right, and are called *coretti* (little choirs). I have a very good picture of the left *coretto*, which the *Museo civico* at Padua gave me many years ago. The following geometric-perspective analysis is based on this one photo (see fig. 7).

Above a parapet panel a Gothic arcade reveals a view into a room with a Gothic ribbed vault that is apparently supported by four narrow corner columns and from whose boss a large chandelier hangs by a rope. The back wall displays a pointed arch like the front arcade and opens up into a window with a slender, square-edged central column. The left side wall is overlapped by a rounded arch and is thus lower than the back wall. The front arcade consists of two short piers that nestle into the pilasters of the Syrian arch and support the pointed arch. Its front face is smooth but bordered by a profile molding.

Fig. 7. Arena Chapel, *coretto* and static *costruzione legittima* (Underlying photo reproduced courtesy of the Museo civico, Padua)

Let us examine this molding more closely (fig. 8). It is strong and sculpted on the left side (where a subtle intersection with the pilaster occurs), whereas on the right it becomes much narrower and finally turns into just a line (fig. 8, right). The molding is therefore represented in perspective and that implies the finest observations of real constructed architecture, which is expertly simulated as real here. The astonishing perspectival realism of this detail now arouses interest in the overall perspectival-geometric design of the *coretto* (fig. 7). My analytical tool is the *costruzione legittima*, which I place like a template on the painted *coretto* to measure its construction.

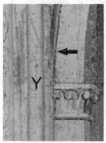 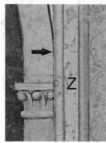

Fig. 8. Details from fig. 7

The first step is to reduce the image field, whose one side must be equal to the side of the square to be foreshortened. Because the *coretto* is placed between the large pilasters, which have been painted according to another perspective construction, it is sensible to divide the width of the entire front arcade, that is, the line segment YZ, which is then at the same time the length of the side of the basic square. I look for the lower boundary of the image field at the foot of the parapet panel on line CD, the top boundary at the crown of the arch (X) of the inner line of the profile surrounding the pointed arch, because this line still just barely exists at Y and Z. The image field is thus rectangular and defined by the vertices ABCD. (Whether these boundaries are correct or not will be proved by the later reconstruction of the perspective design.)

Now we must find the vanishing point (F), that is, the correct position of the eye, from which the *coretto*, if it were real, could be seen from F exactly as it appears on the picture. But because we are looking into the flank of the left pilaster but not into that of the right pilaster, it seems likely that the vanishing point is to be found along the vertical BC line. Assuming (correctly) that the capitals of the piers in the front arcade lie at the same height as the capital (E) of the slender column in the left corner of the *coretto*, I draw a straight line from Y over the abacus of that capital (E) (fig. 9b) to the right until it intersects BC at F1, the vanishing point. What is special about this is that F lies on one side (BC) of the frame of the image, so that the "visual ray" FC thus forms one side of the visual triangle, making a differentiation between a "visual ray method" and a "distance point construction" (Vignola's *due regole*) unnecessary and impossible. If in fig. 3a one imagines the vanishing point A in place of C, then one sees the perspective construction of the *coretto*. To create a perspective according to the *costruzione legittima* we would need the distance point; for the present we do not need it, because we are simply analyzing an existing perspective construction. Without deviating in the least from the rules of the *costruzione legittima* we now lay three planes in the image field ABCD (fig. 7), namely through AB, YZ and CD, which as trapezoids each represent the basic square (a) with the side length CD.

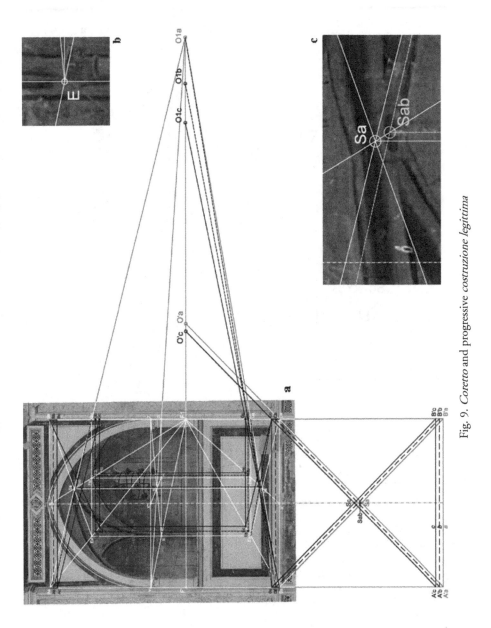

Fig. 9. *Coretto* and progressive *costruzione legittima*

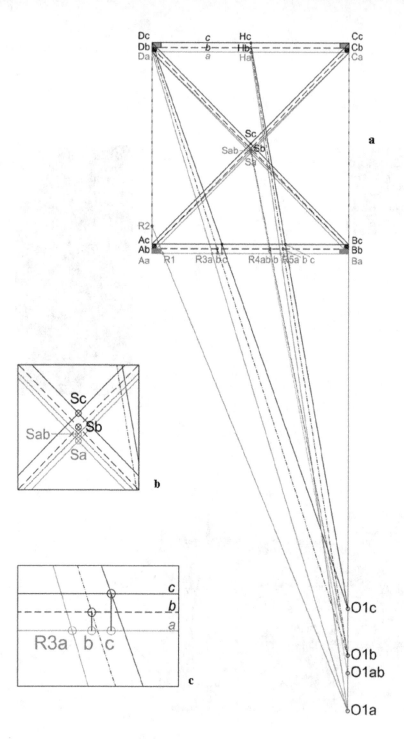

Fig. 10. *Coretto.* Reconstructed ground plan with progressive *costruzione legittima*

The line parallel to YZ through E intersects the perpendicular BC at G; YZGE is the deformed basic square (a) represented as a trapezoid. In the trapezoid one can draw in the diagonals or the central bisecting line to determine the midpoint S. The central axis XX' intersects YZ at X'', and the vanishing point through X'' (X''F1) intersects EG at H. The planes through AB and CD are constructed in the same fashion. As fig. 9 shows, the projections of the line segments AG'', YG and DG' then intersect the horizon at O1, the distance point.

We now remind ourselves that the *costruzione legittima*, i.e., the geometric simulation of the visual image, always involves the perspectival deformation of a square. If so, however, in the image shown in fig. 7 the intersection point S'' must fall exactly in the middle of the boss of the ribbed vault, and the crown of the arch on the back wall of the *coretto* must lie exactly on H'' (and not just to the right of it) and the perpendicular from H'' must coincide closely with the window column. But this is not the case, and the hypothesis that the perspective of the *coretti* is based on the *costruzione legittima* does not appear to be confirmed.

The problem we are presented with here will be better understood when we take a look at the picture of the reconstructed ground plan of the *coretto* (fig. 10). It shows that its perspectively relevant components cannot fit into one square, but rather are distributed among three squares: the basic square (a) and two other squares identified as (b) and (c). These three squares are partially displaced against each other in the same direction – by the value of half of the width of a pier in the front arcade – so that together they form a rectangle. This is the solution to the problem of the perspectively *legittimo* deformation of a rectangle as well: It is divided up into individual squares, which then, each one for itself, is foreshortened according to the rules of the *costruzione legittima*. Three didactic drawings should help to make the method clear (fig. 11a-c).

In fig. 11a, ABCD represents the basic square that is to be deformed, ADF the visual triangle (= two-dimensional visual pyramid), F the vanishing point, O1 the distance point and FO1 the horizon. The straight line AO1 intersects BF at C', the parallel line to AB through C' intersects AF at D'. The trapezoid ABC'D' is then the perspectively deformed square ABCD, to which we now attach a second square at AD with the vertices A' and A''. The straight line A'O1 intersects AF at D' and BF at C''. D'C'C''D'' are now the vertices of the trapezoid that has resulted from the deformation of the square A'ADA''. Both squares together appear deformed as the trapezoid ABC''D'' [Vignola 1583: 68-69]. If one connects the vertices by diagonals, the geometric center of the trapezoid is determined by their point of intersection and the extension of AC'' intersects the horizon FO1 at O2. Such a "wandering distance point" would be mapped to each further square that was to be foreshortened, i.e., square 3 → O3, square 4 → O4, etc. In reality this means that the double square, seen from the distance FO1, as well as the basic square ABCD, seen from the distance FO2, both appear as the same trapezoid ABC''D''. This construction is *legittimo* because the perspectival deformation of squares lined up adjacently takes place in the same visual triangle, that is, in the same visual pyramid, whose apex lies at F or O1. The "wandering distance points" are not needed for this construction, but they can be applied in the simulation of a basic square mutated to a rectangle (see fig. 11c). This construction of the foreshortening of a series of squares was first described by Serlio and Vignola [Vignola 1583].

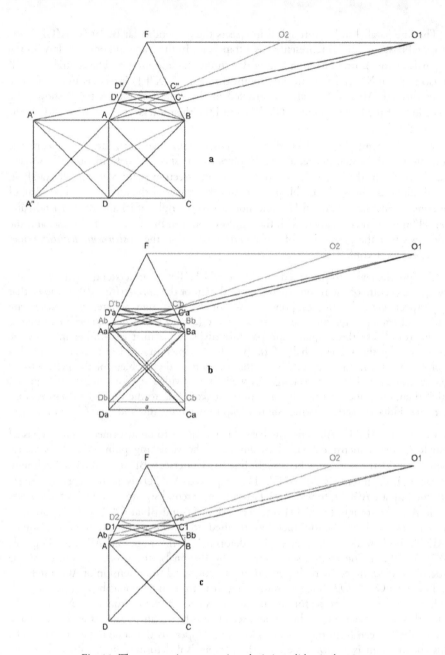

Fig. 11. The progressive *costruzione legittima*: didactic drawings

Fig. 11b and fig. 6 show that not only series of squares, but also congruent squares, partially displaced against each other in the same direction, (that is, together forming a rectangle) can be *legittimo* deformed. Connecting Aa with O1 results in the trapezoid AaBaC'aD'a at the points of intersection with the visual triangle; connecting Ab with O1 results in D'aC'aC'bD'b. The trapezoid AaBaC'bD'b represents the rectangle AbBbCaDa. The construction principle is that the *costruzione legittima* is applied separately to each of the two squares, but these must be mutually interlocked with each other. This is also seen at the points of intersection of the diagonals; (a) and (b) have separate midpoints, but they both share the gray point in between: the point of intersection of the diagonals of the rectangle. The same applies to the trapezoid. – Whether this construction has already been described before, I don't know. Fig. 11b is at the same time the model of the perspective construction of the *coretti* of the Arena Chapel. Fig. 6 demonstrates the construction on the perspective apparatus; the gray lines simulate the displacement of the square and the geometrical consequences.

Fig. 11c shows how an arbitrarily selected "wandering distance point", here O2, can help determine the extent of the displacement of the square. AO2 intersects BF at C2; the extension of O1C2 intersects the extension of DA at Ab. The line segment AAb represents the displacement.

Fig. 3 represents the two methods (Vignola's *due regole*) of the *costruzione legittima*, fig. 11 their two variants, which I would like to call static and progressive. I call static the deformation of the individual square; progressive a series of squares or their partial and monoaxial displacement to the figure of a rectangle.

Now we can better understand the ground plan of fig. 10a. The squares (a), (b) and (c) have been displaced against each other exactly by the value corresponding to half the width of the pillars of the front arcade; yes, the cross-sections of these piers themselves could be considered rectangular in the ratio of 2:1 (which can be easily measured and calculated). The round columns have the same ratio, one of which can be seen in the back corner of the *coretto*, the diameter of half the width of the pier. Hence the piers stand at the bottom between square (a) and square (b) and at the top (as suggested by Giotto) between square (b) and square (c); the columns stand accordingly between square (b) and (c) as well as between square (a) and (b). Their diagonals are drawn in with the points of intersection of Sa, Sb and Sc. Point Sab results from the interlocking of squares (a) and (b) through the diagonals AbCa and BbDa (fig. 10b). The ribs of the Gothic vault lie between AaCc and AcCa as well as BaDc and BcDa.

We now transfer the "displaced squares" to the "image" of the *coretto* (fig. 9a and fig. 12). Below the horizon the squares are pushed upwards by CaDa (Da, Db, Dc); above the horizon they are pulled down by AaBa (Aa, Ab, Ac). The straight lines of Db and Dc to the distance point O1a intersect the perpendicular line CaF1 at Gb and Gc, and the horizontals of these points intersect DaF1 at Eb and Ec. Squares (b) and (c) are also interlocked with square (a). The trapezoid DaCaGcEc is thus the *legittimo* foreshortened rectangle A'aB'aGcDc. If one connects Da with Gb and Gc, the extensions of these straight lines intersect the horizon (through F1) at O1b and O1c: These are the "wandering distance points" of the interlocked squares deformed into trapezoids.

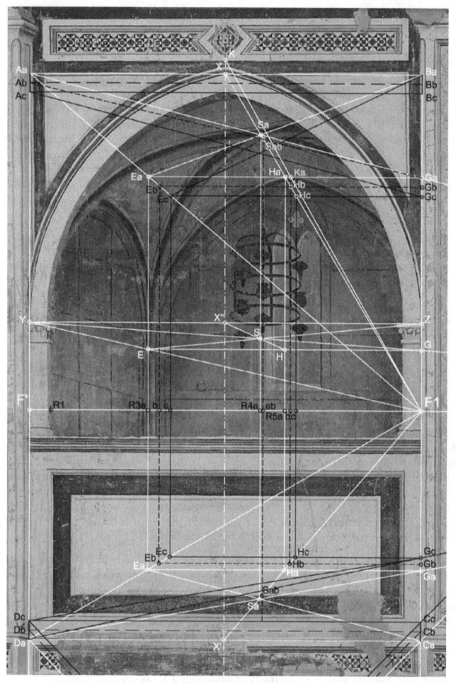

Fig. 12. *Coretto* and progressive *costruzione legittima*

These same operations are performed in the upper plane through AaBa. The straight line AbGa intersects F1X at Sab, and this point thus lies exactly at the point of intersection of the diagonals of the interlocked squares (a) and (b), that is, the rectangle AaBaCbDb (figs. 10a and 10b). In addition, Sab lies exactly along the rope holding the chandelier that hangs from the perforated boss of the ribbed vault (fig. 9c, fig. 12). The ground plan fig. 10a shows the interior organization of the *coretto*: The side of square (a) lies in front of the corner columns, (b) on the back wall, and in the center of (c) stands the window column. The three deeper planes – a sequence of planes stacked behind each other – of the perspective construction are the reason why the crown of the back wall arcade and the window column do not coincide with the perpendicular line through H. As shown in fig. 12, they should instead lay on the points of intersection of F1X with EbGb and EcGc, that is, Hb and Hc. The perpendicular of Hc falls indeed through the central axis of the window column, whereas Hb lies too deep, and the real crown (Ka) can not be determined until the perpendicular line intersects Hb with EaGa. The explanation for this apparent violation of the *costruzione legittima* can be found in the fact that the plane AaBaGaEa lies at the level of the boss of the ribbed vault; the crown Ka, however, represents the height of the vault caps, which appear to be supported by the sturdy ribs. The extension of F1Ka intersects the central perpendicular line through X at K; the line segment KX then corresponds to the "real" thickness of the ribs.

The distance points O1a, O1b and O1c on fig. 9a can be obtained by linearly connecting Da with the trapezoid vertices Ga, Gb and Gc (that is the foreshortened squares (a), (b), and (c) interlocked with each other) and projecting these straight lines to the horizon. The distance points are then transferred proportionally to the ground plan fig. 10a, as well as the distance point O1ab. In fig. 9a this point is obtained by laying a straight line from Aa through Sab, allowing its extension to intersect the horizon (not marked in fig. 9a; only in fig. 10a). Each of the distance points belongs to one of the squares (a), (b), and (c); only O1ab has already been created by interlocking square (a) with square (b).

Let us now compare the ground plan fig. 10a with the geometric construction in fig. 12. In fig. 10a the striking points of the perspective from fig. 12 with the related distance points are connected with each other through straight lines, for example, Da with O1a, Db with O1b and Dc with O1c.

The points of intersection of the straight lines with the bottom sides of the square are marked with a point on each of the corresponding sides a, b, c. If one wants to know at what location these points appear in fig. 12, one must first orthogonally project them to the line segment AaBa (fig. 10c). Why? We must imagine the image plane to be perpendicular above AaBa, the "intersection through the visual pyramid", on which all points that should be seen must be present. The progressive *costruzione legittima* works with several planes stacked behind each other, forming the projection surfaces of the squares mapped to them. Each point on these planes stacked behind each other, must then be projected onto the image plane a above the basic square (a), to become at all visible and measurable on the image. The orthogonal projection here leads to the interlocking of the projection surfaces of the squares (b) and (c) with the image plane of square (a). Their points of intersection on AaBa are denoted R, but for reasons of readability, not all the orthogonal projections are shown in fig. 10a. The position of the points of intersection to (a) can be measured equally well on (b) and (c). In fig. 12 the R points are marked on the horizontals F'F1 and admittedly not by transfer of the correct dimensions of those points from fig. 10a, but rather by orthogonal projection of the spatial points, for example, Ea, Sab, and Hc to F'F1. If one now wants to compare the position of the R points in fig. 10a with those in fig. 12, then one divides the length of the line segment F'F1 by that of AaBa, yielding the conversion factor n: One will then see that the positions of the R points coincide exactly in both figures. (The precision of the design and its pictorial execution is much greater than that in Masaccio's *Trinity* fresco.)

That means that the reconstruction of the ground plan (fig. 10a) must be just as correct as the reconstruction of the *costruzione legittima* in fig. 12. Furthermore, that means that we have here a mapped image of this *coretto*. Just as we can establish the real measurements of a body from the copy of it obtained with the help of the perspective apparatus, so, too, can we obtain all the measurements from the painted *coretto*, which would allow us to convert the painting into real architecture, albeit, mind you, Lilliputian architecture. The width AB of the *coretto* is ca. 146 cm, the height of AD 204 cm; the width of the piers in the front arcade is each the 22nd part of AB and amounts to 6.64 cm. The vaulted room with the vertices Ab, Db, Cb, Bb (see fig. 10a) "is" square and "has" a side length AB of 146 cm; the boss of the ribbed vaulting "hovers" at the height AD of 204 cm above the floor. (Here I am still stating the "real" position of the capital of the window column, whereby in fig. 12 I am selecting Da as the origin of an imaginary orthogonal coordinate system: height 176 cm, width 73 cm, depth 152.6 cm.)

The perspective, the "illusionism" of the *coretti* is often seen by art historians, but – as far as I know – has never been precisely geometrically analyzed.[4] By way of example I quote here Roberto Longhi, Samuel Edgerton und Walter Euler.

Longhi wrote:

> *All' effetto di veridica illusione convengono le due volte gotiche concorrendo ad un solo centro che è sull'asse della chiesa e cioè nella profondità 'reale', esistenziale dell'abside ...* [Longhi 1952: 20].

We have seen that each *coretto* has its own perspectival vanishing point. Edgerton wrote:

> On the same triumphal arch wall Giotto painted yet another trompe l'oeil masterpiece, this time entirely in convergent perspective. ... He apparently intended both of them to look as if their painted pointed arch frames continued the rib-vault construction in the illusionist architectural spaces. This may be the first postclassic example of what was to become the most popular perspective tour de force in all subsequent Renaissance painting: the illusion that the frame around the painting is not only fixed in the viewer's actual space but also integral to the imaginary structure depicted in the virtual space [Edgerton 1991: 80-81].

That is correct in principle. However, it is somewhat strange that the scholar, who has devoted himself entirely to researching perspective, in this book (on Giotto's geometry) dedicates only a few pages to the Arena Chapel, only a few lines to the *coretti* and avoids a clarifying geometric analysis; the *coretti* are apparently foreign to him.

Euler wrote:

> *Die Grenzen dieses Illusionismus fallen jedoch bald auf: nicht allein die unpräzis durchführte Linearperspektive – die Tiefenlinien der Wandgliederung verlaufen fast im Sinne der umgekehrten Perspektive –*

sondern überhaupt die Diskrepanz zwischen der Ausdehnung auf der Fläche und der dargestellten Räumlichkeit... [Euler 1967: 68-69].

It is correct that the (imaginary horizontal) stone bands on the walls of the left wall of the *coretto* take a course that is not at all consistent with the perspective construction of the whole, and cause confusion at the same time. But the "reverse perspective", which Giotto liked using in other places in the Arena Chapel (see fig. 13), does not come into play here either. This clearly shows the exactly constructed reversal of the *costruzione legittima* of the *coretto* in fig. 14. If Euler is correct, the stone bands would have to have lain on straight lines ("visual rays"), emanating from the vanishing points F4, F5 or F6 – but they don't do that!

Fig. 13. Arena Chapel. Detail of the arcade of The Last Judgment. "Reverse perspective" (Photo reproduced courtesy of the Museo civico, Padua)

In fact, the stone bands follow an entirely different geometrical construction which has nothing to do with perspective (fig. 15). We draw horizontal straight lines along the lower edge of the horizontal stone bands of the back wall, until they hit the oblique stone bands on the left and denote these endpoints with a, b, c. The next step is to plot the line segment G'Z from G' with the compass to G'O1 at a. After dividing by the square root we obtain the following geometrical series: G'a : $\sqrt{2}$ = ab; ab : $\sqrt{2}$ = bc. If one connects these points (a, b, c) with the a, b, c points in the corner of the *coretto* by a straight line and extend this to the perpendicular line AD, these straight lines run exactly along the bottom edge of the oblique stone bands. We end up with the points of intersection a, b, c on AD, and they are equidistant from each other: ab = bc. As mentioned earlier, the strange construction has nothing to do with the *costruzione legittima*; at first sight it looks like the solution to a difficult geometrical task – but it is actually quite simple. It is enough to fix the points a, b, c equidistantly along AD and G'O1 in the sense of the $\sqrt{2}$ progression. By linearly connecting the points and after arbitrarily intersecting this great number of straight lines to the corner of the *coretto,* the height of the stone bands is produced as a secondary result. It is self-evident that Giotto, instead of orienting these stone bands central perspectively to F, consciously created a striking disruption in the perspective construction. Why? That will remain a resolvable enigma for the time being. In any case these apparently rather awkwardly drawn stone bands have prevented the art and perspective researchers from considering the *coretti* to be worthy of more detailed examination.

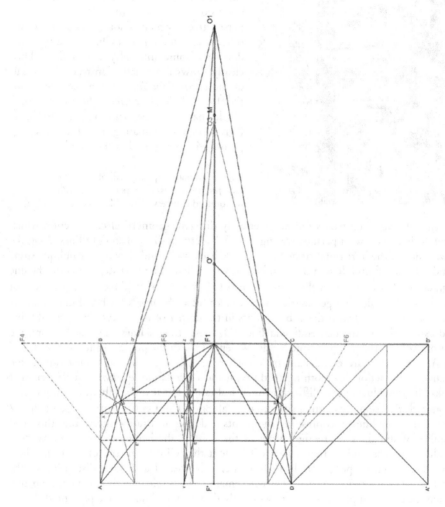

Fig. 14. Static *costruzione legittima* of the *coretto* and construction of "reverse perspective"

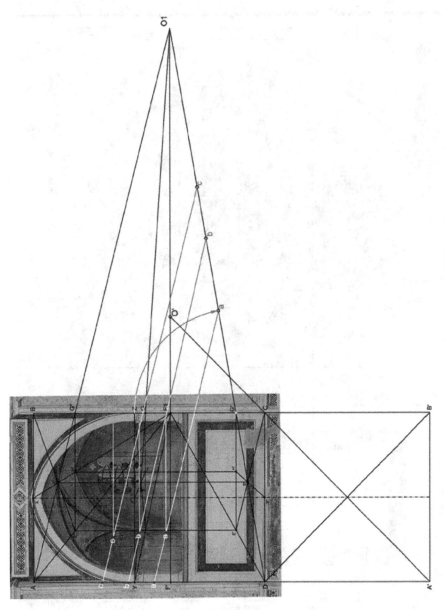

Fig. 15. . Coreto. Geometric construction of the oblique stone bands

Fig. 16. *Coretto*: Static *costruzione legittima* and surface geometry

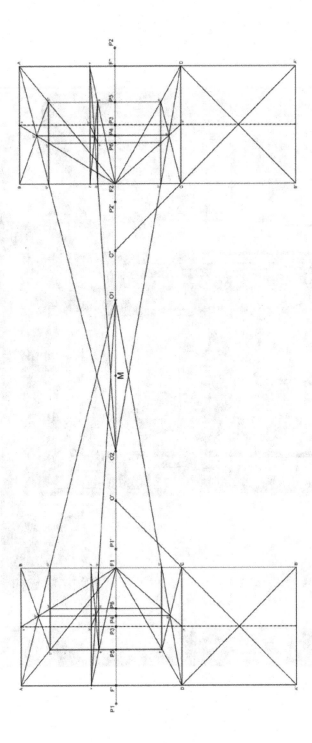

Fig. 17. The reflected static *costruzione legitima* of the two *coretti*

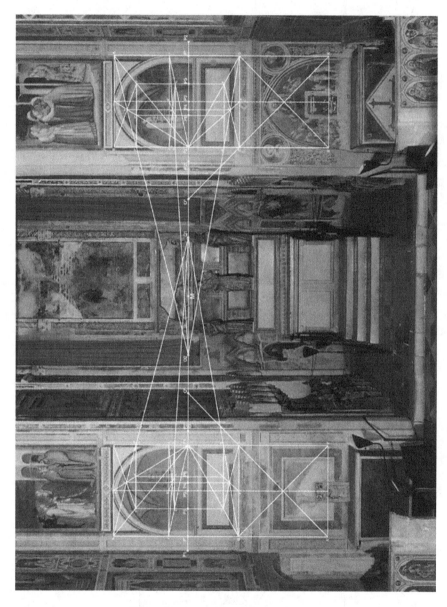

Fig. 18. As fig. 17, projected into a photo of the choir wall of the Arena Chapel

Using surface geometry, independent of perspective construction but still linked to it, Giotto was not sparing in any of the design of the *coretti* – see fig. 16. If one draws the square DCC'D' over DC, then its diagonals intersect at P4, that is, exactly at the height of the parapet. If one connects P1 (the point of intersection of the central perpendicular lines XX' with F'F1) with A and B, then P1A and P1B intersect the horizontal YZ at P2 and P3. These are the exact midpoints of the arc of the front arcade: D'B intersects F1X at P5, that is, the point Sab that resulted from the *costruzione legittima*; P1B intersects CX at P6, that is, at the place where the window column stands in terms of surface geometry; the arc around C with CD passes through E. The analysis could go on like this.

The right *coretto*, so it seems, is the mirror image of the left one and vice versa. By reflecting the left perspectival basic construction on the horizon around the midpoint M, a symmetrical line drawing results (fig. 17). If one projects this drawing, whose left half is totally adapted to the image of the left *coretto* (because it was developed from it), onto a photo of the lower section of the chancel arch (fig. 18) it can be seen that on the right side there is a slight displacement between the "image" and the drawing. This displacement may be due to the photo, which was not taken exactly along the central axis of the chapel; however, it could also be that the choir wall is not entirely symmetrically built. Final clarity can be established only by a photogrammetric photograph. So, for the time being, it must remain open as to whether the perspective constructions of the two *coretti* coincide so exactly with each other that one could speak of a mirror image. Yet this last statement is what I consider a prerequisite for a working hypothesis in the following analysis.

Fig. 17 shows the *costruzione legittima* of the left *coretto* and its reflection. It is striking that quite a number of partial line segments along the line segment P1P2 exhibit the golden ratio (*sectio aurea*) to each other. I measured them with proportional dividers and listed them with the *minor* on the left, the *maior* on the right.

MINOR	*MAIOR*
F'O1	F'F"
P1O'	P1M
P1P1'	P3M = P1'O1 = O'O" = P1O2
P1P6	P1P1'
F1O1	F'O"
F'F1	F1M
P1P3	P3O' = MO'
BZ (=ZG')	YZ

The divisions listed here are subject to the above-mentioned admissibility of the "mirror image", and they were certainly not all intentionally made: many may have been a secondary result. However, I have no doubt that Giotto applied the *sectio aurea* here, as an old and noble proportioning method. I forgo further conjectures and am confident that it will soon be possible to photograph the choir wall of the Arena Chapel photogrammetrically and analyze it more precisely.

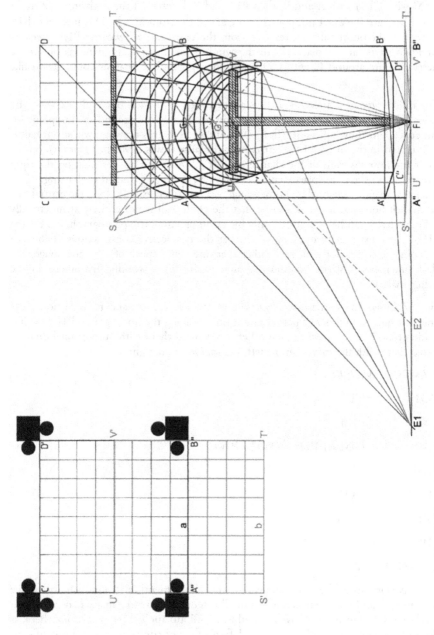

Fig. 19. Masaccio, Trinity fresco: Ground plan and frontal perspective with displacement of basic square

Let me add a comment on the "illusionism" of the *coretti*. Despite elements of surface geometry, they actually represent central perspective: they have been designed and painted according to the *costruzione legittima*, that is, as visual images. Yet the two *coretti* have no common vanishing point and their vanishing points F1 / F2 lie at more than twice the eye level of an observer. Let us image that the *coretti* are constructed architecture: we would then have to climb scaffolding to bring our eye to the position of F1 / F2 and the distance of O1 / O2 to see the constructed *coretti* exactly as Giotto painted them. From the floor of the chapel we would have a different visual image of these constructed *coretti* and, similarly, the painted *coretti* appear distorted to us from the floor. However, this distortion is so slight that it is hardly noticeable. We are well aware that Giotto's *coretti* are simulations of visual images based on the *costruzione legittima*, but these simulated visual images have no natural – that is, no geometric – reference to the observer: vanishing point and eyepoint do no coincide at one point. The term "illusionism" should be avoided here. However, it is striking that Masaccio in his *Trinity* fresco (fig. 1) places the eyepoint at the eye level of a person of normal height (167 cm) and a few years later Alberti in his *Pittura* (1435) makes precisely this – the identification of eyepoint and vanishing point in wall painting – a formal principle.[5] This is the evolutionary step the "Renaissance" took beyond Giotto in the direction of illusionist painting.

Finally, I would like to demonstrate that the method of "progressive *costruzione legittima*" which I have described in the *coretti* was also used by Masaccio in his *Trinity* 130 years later (figs. 1, 19), not to expand the pictorial space to the back, but rather to the front, to the observer, beyond the wall defined as the projection surface. My brief demonstration will not be readily understood by those readers who are not familiar with my Masaccio article [Hoffmann 1996].

Fig. 1 shows the entire fresco, heavily restored in many sections. The points marked F (= vanishing point), A and B (AB = width of the basic square a) coincide with the corresponding points on fig. 19.[6] The frontal perspective is created as follows: The basic square (a) with the vertices ABDC is drawn along the line segment AB and divided into 8 x 8 = 64 squares corresponding to the 64 coffers of the barrel vaulting (see fig. 1). Linearly connecting these panels to AB with F produces the trapezoid ABD'C', through which the coffered barrel vaulting is represented by semicircles. In front of the "wall" and in front of the vaulted canopy with the "Trinity" inside the donor pair kneels on a platform, which lies just slightly above F. The question is then, what "real" depth must this platform have to provide room for the donor pair? Because the space above the basic square (a) is already filled with the figures of the Holy Trinity, we must create additional perspective space (fig. 19, left) by displacing the square a to the front (and then calling it b). On the frontal perspective (fig. 19, right) we (mentally) displace the a-square (= ABDC) upwards by half its width, so that its base lies on I. The extensions of the horizontal lines through I and the orthogonals FA / FB intersect at S and T. ABTS is then the half basic square (a) deformed into a trapezoid, which – projected onto the legs of the visual triangle FT'S' – yields the donor platform with the vertices S'T'B"A". Because the side AB of the real basic square is 210.5 cm long, the "real" depth of the platform is then 105.3 cm. The life-sized donor figures thus have enough room to kneel. On the frontal perspective it can readily be seen that the expansion of the space gained by the displacement of the square will also always produce a homogeneous perspective of the whole space.

Conclusion

I believe I have proved that the *coretti* were designed and painted according to the rules of the progressive *costruzione legittima* and thus represent simulations of visual images. Because no images of this type have come down from Classical Antiquity and because the literary references remain silent in this respect – as Berthold Hub demonstrated recently in a comprehensive study [2008] –, the *coretti* must be considered, according to today's knowledge, as the oldest monuments manifesting the application of the *costruzione legittima*; that means the history of the central (linear) perspective must be rewritten. In any case it was not a Renaissance invention. I expressly agree with the researchers[7] who see Giotto's painting in conjunction with the findings of the Scholastic "optics specialists" (such as Grosseteste, Witelo, Bacon), all of whom stood with their feet firmly planted on the ground of Euclid's rigidly geometrically conceived visual theory and its Arab commentators.

Proof is still missing, unless one would allow the *coretti* to constitute the proof of Giotto's intimate knowledge of the Euclidean-scholastic visual theories. I have distinguished between the image producing perspective (simulating the visual image) and the copying perspective (fixing the visual image). The invention of the perspective apparatus by the young Brunelleschi (ca. 1400), in which the cross-section through the visual pyramid was laid by means of the grid of threads, thus yielding the perspectival copy of a real body also assumes knowledge of the Euclidean visual theory. Looking at it like this, it also seems to me that this invention fits better to the years around 1300 than to the later years around 1400. That must be clarified more precisely.

Acknowledgment

Dipl.-Ing. Nikolaos Theocharis (Institut für Kunstgeschichte der Universität Bern) was my research assistant from 2001-2004 in the reconstruction of the geometric design of Hagia Sophia in Istanbul and also provided invaluable assistance in editing this essay on Giotto's *coretti*; the results of this examination could not have been obtained nor presented without his accurate AutoCAD drawings. I am very grateful to him for this as well as to Prof. Dr. Bernd Nicolai (Lehrstuhl für Architekturgeschichte und Denkmalpflege der Universität Bern) for his kind support of this work. My thanks are also extended to Birgit Wörz (Institut für Kunstgeschichte der Universität Würzburg) for important photographic work and Karen Christenson for the translation of my text into English.

Notes

1. Now the clearest explanation of Alberti's method is that given by Tomás García-Salgado [1998].
2. Cf. [Vignola 1583], in particular on p. 18 the ingenious figure under *teorema terza*, depicting the *due regole* and proving at the same time that they lead to the same result; Vignola-Danti also mention that Serlio had already described but incorrectly depicted the *due regole*, a mistake, *il quale nasca dalla stampa*. See also [Sinisgalli 1978: 62-63].
3. H. Wieleitner [1920], for example, would like to see the term reserved for the "visual ray construction" (Vignola's *regola prima*); see [Hoffmann 1996: 76]. See also [Grayson 1964] and [Parronchi 1964].
4. The most important writings on this are mentioned in [Kohnen 2004].
5. Alberti, Della *Pittura*, Libro Primo, 19: "*...però che così e chi vede e le dipinte cose vedute paiono medesimo in suo uno piano*". Tomás García-Salgado particularly emphasized the significance of this passage and correctly interpreted it as follows: "The observer and the central vanishing point are the extreme points of the symmetrical line of sight: In the Albertian model, the central vanishing point is given by the central point, which must correspond to the

observer's eye level so that what the observer paints is on the same floor as the painter" [García-Salgado 1998: 120].

6. The vertices C' and D' in fig. 1 lie a little higher than in fig. 19, because when Masaccio painted his frescoes he deviated by 4.5 cm from his design. The black lines in fig. 19 show the reconstruction of this former design. This is of no importance here for the interesting question of "displacement of the square".

7. Specifically, see [Bergdolt 2007]. In general, see [Lindberg 1976].

References

ALBERTI, Leon Battista. 2000. *Das Standbild – Die Malkunst – Grundlagen der Malerei.* Herausgegeben, eingeleitet, übersetzt und kommentiert von Oskar Bätschmann und Christoph Schäublin unter Mitarbeit von Kristine Patz. Darmstadt: Wissenschaftliche Buchgesellschaft.

BERGDOLT, Klaus. 2007. *Das Auge und die Theologie: Naturwissenschaften und "Perspectiva" an der päpstlichen Kurie von Viterbo - ca. 1260-1285.* Paderborn: Ferdinand Schöning.

EDGERTON, Samuel Y. 1991. *The Heritage of Giotto's Geometry: Art and Science on the Eve of the Scientific Revolution.* Ithaca and London: Cornell University Press.

EULER, Walter. 1967. *Die Architekturdarstellung in der Arena-Kapelle. Ihre Bedeutung für das Bild Giottos.* Bern: Francke Verlag.

GRAYSON, Cecil. 1964. L. B. Albertis "costruzione legittima". *Italian Studies XXIII:* 14-27.

HUB, Berthold. 2008. *Die Perspektive der Antike. Archäologie einer symbolischen Form* (Europäische Hochschulschriften, Reihe XX Philosophie, Bd. 720). Frankfurt am Main: Peter Lang.

GARCÍA-SALGADO, Tomás. 1998. Geometric Interpretation of the Albertian Model. *Leonardo* **31**, No. 2: 119-123.

HOFFMANN, Volker. 1990-1992. *Filippo Brunelleschi: Kuppelbau und Perspektive.* In: Quaderni dell'Istituto di Storia dell'Architettura, 15-20, (=Saggi in onore di Renato Bonelli, Vol. I): 317-326.

————. 1996. *Masaccios Trinitätsfresko: Die Perspektivkonstruktion und ihr Entwurfsverfahren.* In: Mitteilungen des kunsthistorischen Institutes in Florenz, Vol. XL, No. 1/2: 42-77.

————. 2002. Messbilder. In: *UNIPRESS, Forschung und Wissenschaft an der Universität Bern* **115**: 27-30.

KOHNEN, Michael. 2004. Die coretti der Arena-Kapelle zu Padua und die ornamentale Wanddekoration um 1300. *Mitteilungen des kunsthistorischen Institutes in Florenz* **XLVIII**: 417-423.

LINDBERG, David C. 1976. *Theories of Vision from Al-Kindi to Kepler.* Chicago and London: University of Chicago Press.

LONGHI, Roberto. 1952. Giotto spazioso. *Paragone* **XXXI**: 18-24.

PANOFSKY, Erwin. 1915. Das perspektivische Verfahren Leone Battista Albertis. *Kunstchronik,* NV, **XXVI**, 41/42: cols. 505-516.

PARRONCHI, Alessandro. 1964. La "costruzione legittima" è uguale alla "costruzione con punti di distanza". *Rinascimento* **XV**: 35-40.

SINISGALLI, Rocco. 1978. *Il contributo di Simon Stevin allo sviluppo scientifico della prospettiva artificiale.* Rome: L'Erma di Bretschneider.

————. 2006. *Il nuovo De Pictura di Leon Battista Alberti.* Rome: Edizioni Kappa.

VAGNETTI, Luigi. 1979. *De Naturali et Artificiali Perspectiva. Bibliografia ragionata delle fonti teoriche e delle ricerche di storia della prospettiva. Contributo alla formazione della conoscenza di un'idea razionale, nei sui sviluppi da Euclide a Gaspard Monge.* Florence: *Studi e documenti di Architettura* **9/10**.

VELTMAN, Kim. 1986. Literature on Perspective: A Select Bibliography. *Marburger Jahrbuch für Kunstwissenschaft* **21**: 135-207.

VIGNOLA, Giacomo Barozzi da. 1583. *Le due regole della prospettiva pratica, con i commentarij del R.P.M. Egnatio Danti dell'ordine de Predicatori. Matematico dello Studio di Bologna.* Rome: Francesco Zanetti.

About the author

Volker Hoffmann studied art history, archaeology and philosophy at the universities of Hamburg, Munich, Rome and Paris. He held an associate professorship in Darmstadt and Würzburg and a full professorship in Munich. In 1991 he was appointed to a full professorship for architectural history and historic preservation at the University of Bern and has been professor emeritus since 2005. His teaching ranged from European architecture as a whole with forays into Byzantine and Islamic architectural history. His research focuses on early Christian, early medieval and Renaissance architecture as well as image theory and design history

Celestino Soddu

Politecnico di Milano
Piazza Leonardo da Vinci 32
20133 Milan ITALY
Celestino.soddu@generativeart.com

Research

Keywords: design theory, didactics,
perspective, representations of
architecture, algorithms, descriptive
geometry, dimensions (first, second,
third, fourth, etc.), dynamics,
Euclidean geometry, infinity,
morphology, non-Euclidean
geometry, non-linear dynamics,
perspective geometry, philosophy,
projective geometry, topology,
transformations, virtual architecture,
virtual reality

Perspective, a Visionary Process: The Main Generative Road for Crossing Dimensions

Abstract. Perspective is the only tool able to create subjective links between human beings and art spaces. Each perspective representation can be drawn only by identifying a subjective point of view, which varies from individual to individual. This represents a subjective way of interpreting the hidden logical structure of the world. By interpreting the "perspective" representations of the artists of our past we can approach their cultural visions through a possible re-constructing of their represented spaces, a non-linear process that opens a creative-generative path from past to future, teaching us that "dolce" perspective is the main road for our logical interpretation of art. Starting in 1979 I designed original software for space representations. My investigations concerned perspective and, in general, representations crossing multiple dimensions.

1 Visionary crossing by moving from a dimension to another

The field of investigation of this paper is the relationship between the three-dimensional form and its two-dimensional image in its manifold variations. But we could consider also the image and its possible 2D forms, in its manifold interpretative variations. The "generative" reciprocity between the form and the image of the form, where every form "produces" a plurality of images and where each image "produces" a plurality of forms in an endless spiral, is one of the principal fields of construction of Generative Art. This art was born from expressing ideas as a morphogenetic logical process.

First of all, a difference of dimension can exist between the form and its image. Often this difference consists in considering the form as a three-dimension event and its image as a two-dimension representation, but this is only one of the possibilities. We can get a 3D representation from an event having many dimensions, or we can increase the dimensions of the representation in comparison to the dimensions of the event, as, for instance, when we try to represent the image of a jewel pending from the neck of a noblewoman in a seventeenth-century portrait by building a three-dimensional object that interprets the image of the painting. In this case only one of the possible two-dimensional representations of the constructed 3D event will fit the original image.

In order for the result of this moving through different dimensions to be considered totally acceptable, it is necessary that each point of the form correspond to one point of the image and that the structure of the form-system will have the same topological logic then the image-system. This is obviously not possible in the passage from one dimension to another. The *perspectiva artificialis* of Piero della Francesca is only one of the possible two-dimensional representations of three-dimensional events. With this approach much information is lost. The inverse run, from the perspective representation to the three-dimensional event is, in fact, only a reasonable hypothesis. This passage can be considered as acceptable only if we build this three-dimensional event on the basis of important additional knowledge that we cannot find in the image, i.e., the location of the point of view used in the representation. If we don't know this, it can only be identified through a subjective interpretation; every interpretation therefore "produces" different forms.

Further, we can reconstruct only what we see and not what is behind or inside the represented events. As Florenskij said, the perspective image represents only the skin of the three-dimensional event linking the three-dimensional event to the two-dimensional representation. But, even in light of this consideration, the bending of the skin will never be sufficiently represented in the 2D sheet of the sketch. The relationship between bending of the skin and the 2D sheet is comparable to the relationships between Euclidean geometry and not-Euclidean geometries.

But we also have to perform a further interpretation choosing among the different techniques of perspective representation that we presume could have been used to produce the two-dimensional images. These techniques are manifold and can be synthesized into three types, each of which can link the form to its image in a different way.

2 Perspective tools

2.1 Perspective – 1 to 1. One point of view to one target

Starting Dimension: 3 (x, y, z of object and viewer)
Representation Dimension: 2 (x, y)
Each bundle of parallel straight lines converges into one point representing infinity

Perspective with only one point of view and only one direction of the gaze. The observer and the represented event are in front of one another and the interface is a plane screen.

This is the *perspectiva artificialis* of Piero della Francesca: only a single point of view (and therefore only one eye and not two), and only a single direction are considered. This direction also becomes the central vanishing point in the geometric construction of the image.

In this case the ambiguities of the correspondence that must be clarified as we move back from the 2D representation to the 3D even are:

1. The location of the point of view. Moving the location higher and lower in the 3D scene changes the horizontal order; for example, the floor will be sloped to a greater or lesser degree.

2. The distance of the point of view from the scene. In the image by Piero della Francesca (fig. 1a), the 3D reconstruction of the image results in a very long space. This is because the distance used in the representation is not congruent with the first impression and with the hypothesis that the floor is composed of square elements, as can be seen by looking at the details.

Fig. 1a. Piero della Francesca, *Flagellazione*. A reconstruction of the space represented results in a very long space, very different from what might be expected

Fig. 1b. Image of a medieval city by Simone Martini. It seems to be a *perspectiva artificialis* but the vanishing point moves across the image referring to a further dimension: it represents the dynamic moving of the observer

These parameters, together with other ambiguities inside the structure of the 2D image, make it possible to arrive at an endless number of different interpretations of the image when we can try to reconstruct the 3D object. This is one of the main fields where it's possible to "generate" many different 3D objects starting from a 2D representation. This because we need to use our "interpretation"; we need to go through a "creative generative logical process".

Further, starting from Simone Martini's depictions of Italian medieval cities (fig. 1b) it was possible to generate endless variations of these cities while maintaining their unique identities (fig. 2). This work, which I developed and published in 1986 and 1989, was based on the possibility to identify, in each image of Simone Martini and Giotto, many stratified perspective representations belonging to the movement of the point of view. Often this movement goes from exterior to interior of these cities generating a complex representation that can be interpreted as 2D representation of 4D events, interpreting as further dimension the sliding position of the observer.

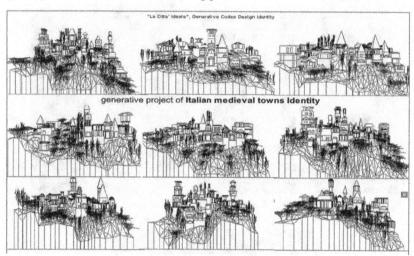

Fig. 2. 3D models of medieval Italian cities generated by the author in 1988. This project was based on the perspective interpretation of Simone Martini's images of cities

2.2 Perspective – 1 to Infinity

One point of view to endless targets.

Starting Dimension: X, Y and Z of object and viewer. There are also the polar coordinates X and Y of targets referred to the motion of sight all around the viewer.

Representation Dimension: 2 (x, y). The 2D representation can be done on a Euclidean (2D sheet) or a non-Euclidean (sphere) interface.

Each bundle of parallel straight lines converges into two points (non-Euclidean geometry)

Spherical total perspective: this perspective technique considers only a single point of view but manifold directions of sight, covering up to 360° in the horizontal (cylindrical perspective with heights in *perspectiva artificialis*) as well as in the vertical (spherical perspective). The observer is at the center of the system.

As we know, in *perspectiva artificialis* only the represented point that is intersected by the direction of sight is not distorted. Every other point of the perspective image is distorted compared to the "view". The distortion is proportional to the distance of the represented point from the point where the gaze crosses the sheet. Spherical total perspective cancels all these distortions. It is constructed by using a sphere as the interface and tracing all the points of the perspective image with the intersection of the gaze in the spherical interface. In this way only undistorted points are used.

After that we have the problem of how to represent this spherical total perspective on a 2D sheet. This can be done by projecting the spherical image onto a sheet rolled into a cylinder. For checking and representing the heights on a non-infinite sheet I use a logarithmic scale. The result is closer to our vision.

Fig. 3a-b. *Generated Castle* by the author represented in total spherical non-Euclidean perspective in two different views, the first one with horizontal sight (above) and the second inclined (below). Images produced using the software designed by the author.

In any case the curved perspectives seem to follow the naturalness of our vision. In fact, if we are inside a space, for instance inside a rectangular room with parallel walls and with a flat ceiling, and we look toward a side, we will see that all the parallel sides of the mentally-constructed image converge toward a point (the vanishing point). Then, if we turn and look at the opposite wall, we see that the same lines converge toward another point opposite the first one. Quickly turning our gaze from one side to the other, we realize that these parallel lines converge in two points of the image that we are building in our mind. Only a non-Euclidean geometry system makes it possible for a bundle of parallel straight lines to converge in two points. The amazing thing is that if we pass from a perspective constructed within a Euclidean geometric system to a perspective constructed within a non-Euclidean geometry, such as spherical geometry, the mathematical representation of the transformation – that is, the algorithm that represents the passage from 3D into 2D – becomes very beautiful mathematically, because it makes it possible to represent everything through the measurement of the angle. I began experimenting on these non-Euclidean total perspectives twenty years ago. These experimentations and the algorithms that I wrote to build the software capable of representing "total perspective" form the basis of my generative software. They constitute a generative engine capable of generating endless possible results starting from a single image [Soddu 1987].

Following this approach, I proved that all perspective events – that is all "subjective" geometrical representations of events going from a very large number of dimensions to dimension two, or, vice versa, going from a very large number of dimension to a very small one – are, when we interpret them in order to go back to the original dimension, "generative engines" very well suited to fitting and communicating our subjective view. Their function of generating endless variations expresses our poetics. It is, in other words, a "logical interpretative engine" capable of representing and managing the complexity of our work. We must remember that perspective, which represents a subjective vision, is the only representation capable of representing infinity on a single sheet.

Further, in practice, if we represent an event with *perspectiva artificialis* and reconstruct the 3D event using a total perspective we can generate an endless number of unpredictable results according to our creative vision.

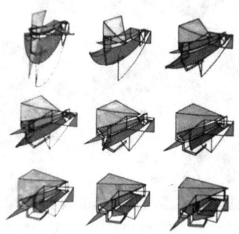

Fig. 4. Moving from Euclidean to non-Euclidean perspective in a reconstruction of the possible 3D objects resulting from an interpretation of a futuristic artwork of Balla [Soddu 1988]

There are some interesting reasons for using this total perspective in architecture. With *perspectiva artificialis* we cannot represent the whole interior space from a point of view inside the space. Thus we cannot control the entire interior system with its relationships and complexity. With total perspective we can do it easily and once we become accustomed to this unusual representation, we discover that this perspective representation is so much closer to our mental approach that we can use it without problems.

There are many different "total" perspectives, depending on the many possible different interfaces that we can use.

The cylinder is the easiest interface, also because it is a "plane" interface curved in one direction only, and allows us to use a normal sheet for our representation, folding it when we want to look at the representation.

We can also use a spherical interface, which is closer to real vision but which requires us to represent a curved surface on a 2D sheet; the alternative is to make our drawing on the surface of a sphere.

There is also a really interesting approach to representing the total environment from its interior. We can use a mirrored cylindrical or conic interface. This system, which we can call anamorphic perspective, was a Flemish and Chinese invention of the sixteenth century (fig. 5).

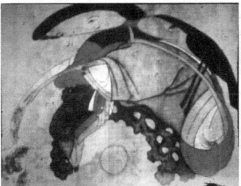

Fig. 5. Chinese anamorphic perspective of two lovers. The circle is the location of the cylindrical mirror

I tried to design the rules for the algorithmic transformation that make it possible to represent in 2D the 3D environment in such a way that it can be viewed by putting a cylindrical mirror in the center of this (circular) drawing (figs. 6, 7, 8, 9).

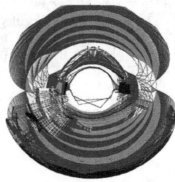

Fig. 6. (above, left). Generated Castle realized with rapid prototyping using 3D STL model directly generated by Argenia, the generative software designed by the author

Fig. 7. (above, right). The Castle represented with anamorphic total perspective using the software designed by the author. The location of the cylindrical mirror is in the middle

2.3. Perspective – Infinity to 1. Infinite points of view to one target

Starting Dimension: X, Y and Z of object and target together with polar X and Y of endless viewers

Representation Dimension: 2 (x, y)

Each bundle of parallel straight lines converges into two points (non-Euclidean geometry)

If we succeed in representing the total interior space with a spherical total perspective, we can think that it is possible to represent a total object from the exterior, looking at the same moment at all the exterior events, not only at the part that we are facing. We could see, together in the same drawing, the front and the rear, the right, the left, above and below. A "primitive" representation of an animal (such as the representations of elephants or lions that I have seen made by people in Somalia) is really similar to this type of perspective. It represents the animal as a carpet-skin and it is possible to look at the top, the right and the left at a single glance.

Fig. 8. Spherical total perspectives of Pantheon in a sequence going from outside to the center. The last perspective is from the center of Pantheon that is the center of the half-sphere of its dome and of its cylinder. This coincidence generates a particular perspective representation. Images made with the author's total/anamorphic perspective software [Soddu 1986]

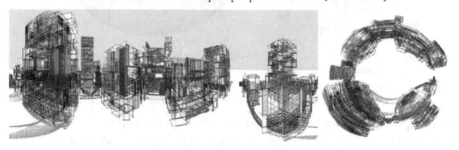

Fig. 9. Image of generated city with spherical and anamorphic perspective realized with the Flemish/Chinese system. In this case, I used, as interface of the anamorphic image, a conic mirror for projecting in the inside of a large cylinder the image of the city, during its generation, viewed by its center. From an exhibition of the author's generative architectures at the Milan Fair in 1998

This is the reverse perspective of Pavel Florenskij [1983]. This approach considers a multiplicity of points of view, the two eyes and their various possible motions, and only one target of the gaze. The represented event is the center of the system. This perspective aims at encompassing the multiplicity of different visions in a single two-dimensional image. This approach tries to represent, in a single 2D drawing, the mental image we form when we look at a 3D object with both eyes, particularly when the object is small and it is very close to our nose. "Reverse" perspective, amplifying this kind of vision, can increase the number of "eyes" up to an infinite number.

The practical construction of this kind of perspective can be created through an interesting conceptual reversal that I have created with my algorithms. If the target of the gaze is unique and the points of view are different we can reverse the total perspective, which has only one point of view and different targets, by setting the point of view in the target and the directions of the gaze in many "eyes". The images thus created could be likened to a representation of the skin of the object seen from the interior but represented as exterior. The reverse perspective has been identified and explained by Pavel Florenskij in relation to Russian icons. Because these are sacred representations, the fundamental choice is setting the represented event as center of manifold views (fig. 10a). In these two-dimensional images the representation of the face of the Saint is, according to my hypothesis, represented as seen from the inside of its head (fig. 10b). Since, as Florenskij affirms, we represent only the "skin" of the physical event we can reverse the face. Its projection on a sheet will turn out to be similar to the representation in reverse perspective of the Russian icons. In other words, it is my belief that the reverse perspective is the reversal of spherical total perspective and not only a reversal of Piero's *perspectiva artificialis*. Also, the Russian icons are only a part of the total reverse perspective, which could represent the back of the Saint's face as well.

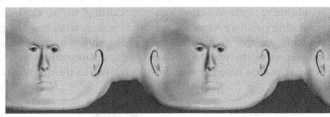

Fig. 10. a, left) Russian icon with Christ represented in "reverse" perspective; b, right) A human head represented in "reverse" perspective. The image is repeated two times (360°+360°) in order to clarify the external representation of the head as a whole. It is made by swapping the 360° interior total perspective, made with the point of view inside the head, into an exterior representation. This swap interchanges the point of view with the target of sight. In the end we have endless points of view and only one sight target. The head is the center of endless subjective points of view. The image is clearly similar to the Russian icons

3 Creative moving across dimensions

The passage from one dimension to another, and particularly from 3D to 2D events through different perspective methods, but above all the reconstruction of the 3D object using different perspective-visual methods introduces fields of variation owing to different factors inherent in the dimensional transformation and in the type of representation used. These fields of variation belong to the subjective interpretation of the image, or better, to the interpretative reconstruction of the parameters that could be used for the production of the image, and of the reconstruction of the parts that are not represented because they are not visible, being either behind or inside the volume of which the skin is represented.

The hypothesis of reading an image by decoding it through the *perspectiva artificialis* when instead it had been constructed using Florenskij's reverse perspective can produce unpredictable forms. For instance a cube could be reconstructed as a pentagonal prism.

This happens because, in reverse perspective, the two opposite sides of a cube are represented as "in sight" along with the side that is facing the observer. The reverse perspective of a cube is capable of showing three faces in sequence because the cube is seen from both left and right. This is what happens when we look at a very small cube and we bring it very close to our eyes: one eye sees the right side, and the other left side; the resulting image is the synthesis of the two views. Our mental image is a reconstruction of the cube representing three consecutive faces. If we look at this representation with a canonical Euclidean perspective approach, we must presume that it is something different from a cube. The space "behind" appears too much ample and the re-constructive interpretation of the 3D form can lead us to imagine more than a single hidden face, for instance two, and therefore to generate an acceptable reconstruction of a prism with five or more consecutive sides. The cube, as a result of these successive passages from dimension to dimension (3D - 2D - 3D) is turned into a pentagonal prism.

These transformations are born of our interpretations: they are a "natural" construction of generative motors that mirror our creative identity, our cultural references.

The idea of an architect isn't based on forms but on transformations. This is a transforming approach that sees the existing world as dynamic, and is capable of generating visionary scenarios and their endless variations. The generative engines are the structure of the designer's idea. They work on morphogenetic codes fitting the uniqueness of the approach; they are the anamorphic logic that allow the designer to generate endless visionary worlds by mirroring, in their multiplicity, the design idea.

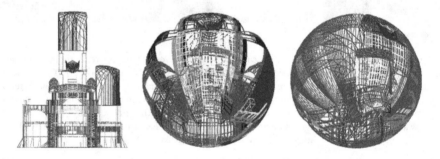

Fig. 11. Generated castle by C. Soddu, represented in elevation and in two different "reverse" perspectives drawn on a sphere, using the software designed by the author for investigating Florenskij's approach

4   Construction of generative morphogenetic processes with perspective approach: identity, subjectivity and variations

The identity of an artwork exists if people can recognize it as belonging to a species. All possible 3D interpretations of a perspective image are recognizable because they belong to the 2D image. In these reconstructed 3D models we can find the particular view that fits the original image; all the other views are different but, at the same time, they are recognizable as belonging to the same species. In reconstructing the 3D space, we constructed an anamorphic object capable of changing according to different subjective views. Each perspective image can be likened to an artificial DNA strand capable of generating a species of 3D events.

When we want to construct the identity of our artworks, we can identify its species and execute it by designing an artificial DNA strand. This approach is Generative Art: building a series of logical rules of transformation capable of generating an endless number of possible results recognizable thanks to the morphogenetic paths used for their creation and to the reference to possible anamorphical rules of logic belonging to our creative and cultural identities.

The results, in terms of quality and extended appreciation, are best where the anamorphic logics produce answers pertinent to different subjectivities, therefore where the generated complex system doesn't provide only the possibility of being understood as axiomatic structure of a shape or of a function but its complexity makes it suitable for subjective and unpredictable uses. This usability is performed and appreciated when the suggestions, the logical rules of use and the aesthetical appreciation of each user are related to the complexity of the designed system and to the potential anamorphic interpretations that this complexity makes possible.

Further, the identity has to belong to a species, not in denial of, but as a reinforcement of the identity as individual, as *unicum*. This leads to the consideration that the design of morphogenetic paths rather than of shapes doesn't take anything away from the final results in terms of identity but rather strengthens them, especially because of the parallel presence of "variations", as occurs in music, from Bach to Mozart and to jazz. Variations are constructed by consolidating different forms at different moments, but these results are reciprocally congruent because of the common morphogenetic paths that, from the detail to the whole, lie at the basis of an idea. These "endless" variations might seem aesthetically less strong and functional, less recognizable than the single result chosen because, at the end of the optimisation of the form-function relationship, it was considered the best, but this approach is misleading. The affiliation to a species, with the possibility of mirroring each result in an infinity of parallel variations, creates two congruent layers of recognizability and identity that are mutually reinforcing: the identity of the species and that of individuals.

The fascination of perspective images are strongly related to the possibility of multiple subjective interpretations: each image is a possible mirror of a different observer's subjectivity.

On the left the rendering of normal perspective; in the center the 360° spherical perspective; on the right the anamorphic perspective with a conic mirror interface. The inclination of the 360° spherical perspectives follows and represents the helical geometric structure of the 3D models.

5 Different perspective views together in the same image

As we can see in the art of Simone Martini, many artists identified in the perspective view a possibility of representing the fourth dimension, that of time. The aim is to achieve the complexity, the quality that belongs to the multi-significance, full-of-sense works of art.

The *Carceri d'invenzione*, the most interesting series of engravings by Piranesi, were made by stratifying, in subsequent moments, further objects and, in the meantime, also further perspective points of view in the same artwork.

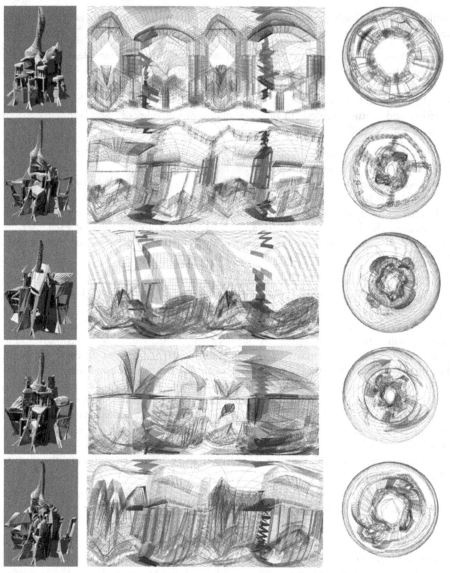

Fig. 12. Variations of the *Babel* project by the author. All these 3D models are completely generated and represented with the author's software

Sergei Eisenstein, in *Oppositions* 11 [1977], describes the increasing complexity of Piranesi's engraves through the addition of new stratified layers. Piranesi used different points of view for these "new" perspectives in a way such that these new layers perform jumps in scale and in space geometries. Eisenstein argues that this method is similar to the vertical sequences in Japanese Kakemono paintings, with the difference that Piranesi unpredictably magnified the subsequent layers instead of reducing them in accordance with the rules of perspective (fig. 13).

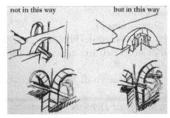 

Fig. 13. Sergej M. Ejzenstein, about Piranesi perspective and Japanese Kakemono [1977]

I tried to go on with this increasing complexity approach, with this creative path, by stratifying into the Piranesi engraves subsequent objects using a perspective point of view that is not exactly the same, but somewhat changed. And I have created this increasing complexity by generating, with my Basilica software, fifty different variations of the same Tower of Babel and using them to create another layer in an engraving of Piranesi, pushing in depth the existing layers. These fifty variations were a gift to the fifty participants of the 2008 Generative Art conference in Milan (fig. 14).

Fig. 14. Fifteen different variations of the Babel Tower, the after-Piranesi generative artwork

In conclusion, Perspective is a "logical interpretation" of reality that is able to generate endless "logical interpretations" fitting different subjective observers.

The availability of different perspective tools creates the possibility to cross dimensions, going from a perspective representation to its 3D space and back to different perspective representations by interpreting the perspective images with our creativity. This possibility is one of the more interesting ways to connect our past, our cultural heritage, to our vision of the future. It increases our creativity.

References

BALTRUSAITIS, Jurgis. 1969. *Anamorfosi o magia artificiale degli effetti meravigliosi.* Milan: Adelphi.

BATTISTI, Eugenio. 1981. *Anamorfosi, evasione e ritorno*. Rome: Officina.

EISENSTEIN, Sergej M. 1977. Piranesi or the fluidity of forms. *Oppositions* 11.

FLORENSKIJ, Pavel. 1983. *La prospettiva rovesciata e altri scritti*. Rome: Casa del libro.

GIOSEFFI, Decio. 1957. *Perspectiva artificialis* Trieste: Istituto di storia dell'arte antica e moderna dell'Università di Trieste.

———. 1999. Introduction to "Logica e Forma". Seminar at the Politecnico di Milano organized by Generative Design Lab.

GOMBRICH, E. H. 1961. *Art and Illusion, a Study in the Psychology of Pictorial Representation*. New York: Princeton University Press.

RAGGHIANTI, Carlo L. 1951. *L'arte e la critica*. Florence: Vallecchi.

———. 1974. Arte, fare e vedere. Florence: Vallecchi.

ROSSI, Paolo Alberto. 1981. Prospettiva invenzione ed uso. *Critica d'Arte* 175-177: 48-74.

———. 1985. La scienza nascosta, analisi delle architetture e pitture del gruppo Brunelleschi & C. Exhibit catalogue. Brescia.

ROSSI, Paolo Alberto and Celestino Soddu. 1986. Il calice di Paolo Uccello uno e senza limite. Critica d'Arte 8: 85-90.

SODDU, Celestino. 1986. *L'immagine non Euclidea*. Rome: Gangemi.

———. 1988. Un'indagine dell'idea di spazio nell'arte contemporanea. *Critica d'Arte* 16.

———. 1989. Città Aleatorie. Milan: Masson publisher.

———. 2004a. Generative Design / Visionary Variations - Morphogenetic processes for Complex Future Identities in the book *Organic Aesthetics and generative methods in Architectural design*, P. Van Looke and Y. Joye, eds. *Communication & Cognition* 36, 3/4.

———. 2004b. 变化多端的建筑生成设计法 (Generative Design). *Architect*, December 2004 (in Chinese).

———. 2005a. Milano, Visionary Variations. Rome: Gangemi.

———. 2005b. Generative Art in Visionary Variations. *Art+Math=X* conference proceedings. Boulder: University of Colorado.

———. 2005c. Visionary Variations in Generative Architectural Design. *Chepos* 3.

———. 2005d. Gencities and Visionary Worlds. In *Generative Art 2005*, proceedings of the International Conference GA2004. Milan: Aleadesign.

SODDU, Celestino and Enrica COLABELLA 1992. Il progetto ambientale di morfogenesi. Progetto Leonardo.

———. 2005. A Univesal Mother Tongue. *Leonardo Electronic Almanac* 13, 8 (August 2005).

TAFURI, Manfredo 1997. *La sfera ed il Labirinto*. Torino: Einaudi.

WITTKOWER, Rudolf. 1992. Idea e immagine. Torino: Einaudi. (Ital. trans. of *Idea and image: studies in the Italian Renaissance*.)

About the author

Celestino Soddu Celestino Soddu obtained his Master's Degree at Università di Roma "La Sapienza" in 1970. In the same year he passed the State Examination for the Register of Professional Architects and Engineers. He has taught Architectural Representation, Composition and Technology in Italian universities since 1971. He is now professor of Architectural Generative Design at the Politecnico di Milano in the Faculty of Engineering-Architecture. In 1997 he founded and is presently director of the Generative Design Lab at the Department of Architecture and Planning of the Politecnico di Milano. Beginning in 1998 he has organized and directed the annual International Generative Art Conference. He has presented his generative projects and artworks in many personal exhibitions, including exhibits at the Hong Kong Museum, Visual Art Centre, MF Gallery in Los Angeles, IDB Cultural Center in Washington DC, the Pacific Design Center, Los Angeles, the Italian Embassy in Beijing, the International Finance Center of Hong Kong and the Commerce Chamber in Milan. He is the author of numerous books in Italian and English. Video interviews and programs for international televisions were created about his architectural research activity in Italy, China, US and Spain. More information can be found in his website http://www.soddu.it

Agnes Verweij

Delft University of
Technology
Faculty of Electrical
Engineering, Mathematics
and Computer Science
Mekelweg 4 HB04.090
2628 CD Delft
THE NETHERLANDS
a.verweij@tudelft.nl

Keywords: seventeenth-
century art; representations
of architecture; didactics;
geometry; lines; mappings;
transformations;
perspective; perspective
geometry

Research

Perspective in a box

Abstract. Perspective is an optional subject for students of some levels in Dutch secondary schools. A proper final task on this subject is the analysis of existing perspective drawings or paintings. This task is sometimes supplemented by a more creative and challenging assignment, that is, the design and construction of a perspective box. A perspective box is an empty box with, on the inner sides, perspective pictures giving a surprising spatial effect when observed through the peephole. The students who take up the challenge are in the first place inspired by the six still existing antique wooden perspective boxes, especially because they were created by Dutch seventeenth-century painters of architecture and interiors. In this article the setup of the perspective in these boxes will be discussed. But for a clear comprehension, we begin by reviewing the principles of linear perspective and their implications for the way perspective images can best be viewed.

Introduction

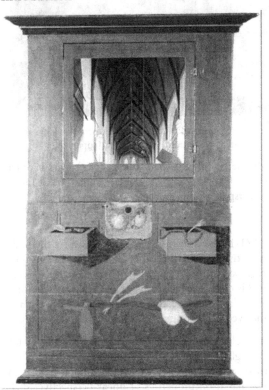

Fig. 1. Anonymous, Perspective box with the interior of a Protestant church, between 1655 and 1660. Front view. © National Museum of Denmark, Copenhagen

Soon after the middle of the seventeenth century, the *perspective boxes* of Dutch painters of architecture and interiors enjoyed a brief period of popularity. These are empty wooden boxes with, inside on the walls, perspective paintings of the interior of a house or a church. Only six perspective boxes still exist,[1] three of which are in the collection of the National Museum of Denmark in Copenhagen. Figure 1 shows one of the two triangular boxes from this museum that display the interior of a Dutch church. The peephole is in the middle of the front panel, decorated with a *trompe-l'oeil* painting. The large square window above it provides the necessary light. When the painting in the box is observed through this window, as in fig. 1, the picture looks odd. For example, the beams of the church are all bent in the middle.

DOI 10.1007/s00004-010-0023-7; *published online* 9 February 2010
© 2011 The Author(s) This article is published with open access at Springerlink.com

Looking through the peephole, the beams are suddenly straight and everything else in the interior fits too (fig. 2). What this figure can not convey properly is the surprising spatial effect that the perspective in the box gives when the picture is observed through the peephole.

Fig. 2. Anonymous, Perspective box with the interior of a Protestant church, between 1655 and 1660. View through the peephole. © National Museum of Denmark, Copenhagen

In this article we will discuss the setup of the perspective in this and other seventeenth-century perspective boxes. But first we will review the foundations of linear perspective and use them to derive how perspective images can best be viewed.

Looking at perspective

The woodcut of Albrecht Dürer (1471-1528) shown in fig. 3 demonstrates the basic principles of displaying in linear perspective. The artist observes a three-dimensional object with one eye from a fixed vantage point through a flat plate of glass, the *picture plane*. His eye is fixed with a tripod in this case. The artist now traces what he sees onto the picture plane. Put differently: he records the intersection points of the picture plane with the (straight) *lines of sight* from the eye to the points of the object to be displayed.

In Dürer's time some construction rules were already known that, especially for straight-lined objects, made the use of a glass plate wholly or partially obsolete. With these rules it was also possible to draw or paint imaginary three-dimensional objects in perspective. The construction rules were based on properties of linear perspective, such as:

- Straight lines that are parallel with the picture plane are displayed as straight lines, while preserving their direction and proportions of distance along the lines;
- Of straight lines that are not parallel with the picture plane the direction and proportions of distance are not preserved; they are displayed as semi-lines of which the end point, the so-called *vanishing point*, depends on the direction of the original line.

Vanishing points are discussed below. Right now our point is not whether a perspective drawing or painting has been constructed or simply created in the manner of fig. 3, but how to observe the result. From this figure can be understood that you see 'depth' in a two-dimensional perspective image when you look with one eye from one specific vantage point. That vantage point is precisely that point at which the artist, in reality or in theory, held his eye.

Fig. 3. Albrecht Dürer. Woodcut in *Underweysung der Messung mit dem Zirckel und Richtscheyt, Fourth Book*. Nürnberg: Hieronymus Andreae, 1525

Unfortunately, perspective paintings in museums are often mounted such that you would have to crouch or stand on a chair or even a ladder to be at the right eye level. And even if the height is correct, it is practically never communicated to the public from which point on that correct eye level they should be observing. Now people today will rarely complain about this; photography, film and television have given them extensive experience in interpreting perspective images seen from 'wrong' angles. But when painting in perspective became an important specialty in the seventeenth-century Netherlands, people naturally did not have that experience yet.

It is unsurprising, then, that in that period the circles of the Dutch painters of architecture and interiors thought up the perspective box as a way to force 'good' viewing in a natural way. With the peephole in the correct place, a hole so small that you cannot look through it with two eyes, things simply work. An additional advantage is that the observer of the images in the perspective box cannot be distracted by his environment. When the images are not only on the wall opposite the peephole, but also on the sections of other walls visible through the peephole, the observer even gets the illusion that he is part of the pictured scene.

Vanishing points in linear perspective

Before we discuss the perspective in the seventeenth-century perspective boxes that still exist, we will review the required knowledge of vanishing points. We assume a situation where an artist uses a glass picture plane to display a line l which is not parallel to the picture plane. We also assume that he uses a sequence of points A_1, A_2, A_3, ... along l and chosen such that their distance to the picture plane grows unbounded. The points where the lines of sight from the artist's eye O to A_1, A_2, A_3, ... intersect the picture plane we will call A_1', A_2', A_3', ... These points lie on the perspective projection l' of l (fig. 4).

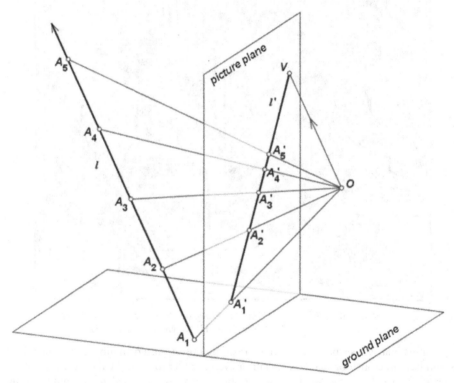

Fig. 4. The vanishing point V of a line l that is not parallel to the picture plane. Drawing by the author

First we note that the angles that the lines of sight OA_1, OA_2, OA_3, ... make with l get smaller and eventually approach 0. So, if n approaches infinity, the direction of OA_n and with it the direction of OA_n' will approach the direction of l.

It follows that the sequence A_1', A_2'. A_3', ... converges to – and that l' therefore ends in – that point V of the picture plane for which OV is parallel to l. This point V is called the *vanishing point of the line l*. Based on what was already stated about the relationship between the eye of the perspective artist and the 'correct' viewpoint of the observer of a perspective image, the result can also be formulated like this.

> **Theorem:** The vanishing point V of a straight line l is the point of intersection between the picture plane and the line of sight drawn from the 'correct' point of view parallel to l.

This theorem is similar to the third theorem asserted and proved by the seventeenth century Flemish mathematician, physicist and engineer Simon Stevin in his treatise *Van de verschaeuwing* [Stevin 1605] to which we return in a subsequent paragraph.

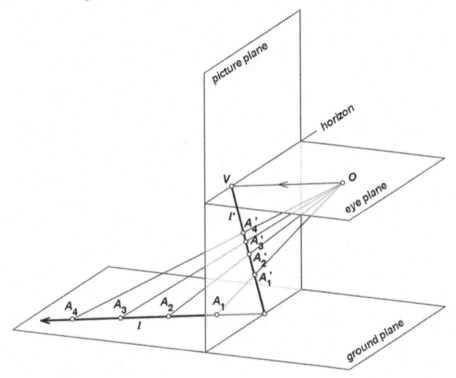

Fig. 5. The vanishing point V of a horizontal line l that is not parallel to the picture plane.
Drawing by the author

If l is a horizontal line, for example a line in the horizontal *ground plane* (fig. 5), then the theorem implies that the vanishing point V of l lies on a horizontal line of sight. This line of sight then lies in the so-called *eye plane*, the horizontal plane through the eye O, and V then lies on the *horizon*, the line of intersection of the eye plane and the picture plane.

Distance points

Now we examine the situation in which the picture plane is vertical, while the object to display in perspective – aside from lines that are parallel to the picture plane – is characterized by lines perpendicular to the picture plane and a number of lines in horizontal or vertical planes that are at a 45° angle to the picture plane. A situation like that is shown in fig. 6, where $ABCD.EFGH$ is a cube. From the theorem above it follows that the line of sight to the vanishing point of the lines perpendicular to the picture plane, is also perpendicular to the picture plane. In the figure these would be the lines AD, BC, EH and FG. So this vanishing point is the perpendicular projection of the eye point O onto the picture plane. This point is called the *central vanishing point*, indicated with the letter P. In these cases the central vanishing point is on the horizon.

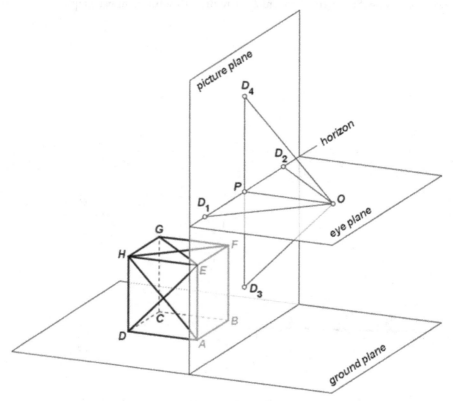

Fig. 6. Vanishing points in the perspective of a cube with edges perpendicular or parallel to the picture plane: the central vanishing point P and the distance points D_1, D_2, D_3 and D_4.
Drawing by the author

Fig. 6 also shows four vanishing points of lines that are at a 45° angle with the picture plane: D_1, D_2, D_3 and D_4. The points D_1 and D_2 are the vanishing points of diagonals of the horizontal sides of the cube, D_1 of BD and FH, D_2 of AC and EG, and therefore lie on the horizon. D_3 and D_4 are the vanishing points of diagonals of vertical sides of the cube. All diagonals of these sides are at 45° angles with the picture plane and with the direction perpendicular to the picture plane. According to the theorem the same then holds for the lines of sight OD_1, OD_2, OD_3 and OD_4. The triangles OPD_1, OPD_2,

OPD_3 and OPD_4 therefore have two 45° angles and one right angle and it then follows that $OP=PD_1=PD_2=PD_3=PD_4$. The *distance* from the eye point O to the picture plane is equal to OP and therefore also equal to the distance between P and the points D_1, D_2, D_3 and D_4. This explains why the points D_1, D_2, D_3 and D_4 are called *distance points*.

When he has recorded the position of his eye O relative to the picture plane, the perspective artist can use this knowledge to determine the central vanishing point P and the distance points D_1, D_2, D_3 and D_4. He can then use these points as vanishing points of the characteristic lines of the object while constructing the perspective image. Anyone who wants to look at perspective drawings or paintings correctly should then look for the vanishing points of those lines that were perpendicular to the picture plane and a vanishing point of lines in a horizontal or a vertical plane that were at a 45° angle with the picture plane. Combining the same knowledge and these points the correct eye point O can be derived.

One-point perspective

The seventeenth-century Dutch painters of architecture and interiors almost certainly did not have the knowledge about vanishing points in the form presented above. They were only familiar with the construction rules for those objects and their positioning relative to the picture plane where the central vanishing point and the distance points on the horizon serve as vanishing points, primarily by the books of examples by Hans Vredeman de Vries. Yet Simon Stevin's *Van de verschaeuwing*, the first thorough treatise on perspective in Dutch (with translations in Latin and French), was published in the same year that Vredeman de Vries' second book was [Vredeman de Vries 1604-1605; Stevin 1605]. However, Stevin's treatise was published in a mathematics book and books like that were unknown to painters. Besides, the perspective theory of Stevin would have been too hard to understand for them because of the mathematical background it required [Andersen 1990; Andersen 2007].

Mostly they used *one-point perspective*, which is perspective in which the characteristic lines are parallel or perpendicular to the picture plane, such that these lines either have no vanishing point, or have the central vanishing point as their vanishing point. Well known examples are the living rooms of Johannes Vermeer (1632-1675) and the churches and church interiors of Pieter Jansz Saenredam (1597-1665) in which the picture plane is always parallel to one of the walls of the displayed building or interior. The distance points on the horizon are merely vanishing points of the diagonals – or, if the tiles are laid diagonally, of the edges – of square floor tiles and sometimes of the edges of a single diagonally placed piece of furniture.

The same holds for the vertical walls of the two surviving rectangular perspective boxes from the seventeenth century. One of these is the box with the interior of a Dutch house of around 1670 which is in the National Museum of Denmark in Copenhagen (fig. 7).

Fig. 7. Anonymous, Perspective box with the interior of a Dutch house, between 1665 and 1675. Front view, without top panel. © National Museum of Denmark, Copenhagen

Of this box, the front panel with the peephole and light opening is lost.[2] The unknown painter has been rather lax with regards to perspective. He has displayed every standing wall of the interior almost completely on a corresponding side of the box. He did, however, paint parts of the tiled floor on the vertical sides of the box. It is interesting that the painting of the tiled floor does not extend to the bottom of the box, while this would have been easy to do. The bottom of the box is actually a horizontal picture plane and therefore parallel with the displayed floor. Therefore the tiles could simply have been displayed similar to their actual form. Maybe the painter in fact has done so, whereas later on for some reason the bottom of the box was replaced by the current one, in which only rectangular carvings indicate the tiling.

For an image of the other rectangular perspective box we refer to the website of the museum that has it: the National Gallery in London.[3] This box was made by Samuel van Hoogstraten (1627-1678), probably between 1655 and 1660, and shows the interior of a Dutch house. The original front panel with the light opening has not survived. This box is special in that it has two peepholes, one in the left and one in the right side wall, at the same height and close to the front of the box. So while constructing the perspective of the left part of the tiled floor on the far wall Van Hoogstraten has had to take into account a different central vanishing point and other distance points on the horizon than for the right part of the floor on the far wall. The painter hid the bad seam between both parts with a round carpet, a chair and a pillow fallen off the chair. The images on the floor and the ceiling of the box presented less of these issues. As explained above, square floor tiles should simply be painted square and the parallel beams on the ceiling stayed parallel in the painting. However, Van Hoogstraten did not leave it at that: he painted parts of (vertical) chair legs and a sitting dog on the floor. But these are only visible from one peephole. Because of this it was always clear which of the two peepholes' perpendicular projection onto the floor of the box should be used as the vanishing point for the vertical lines. Both [Andersen 2007] and [Jensen 2007] have an extensive discussion of the perspective of this box.

Two-point perspective

The books of Vredeman de Vries also show several images in the perspective that forms when the vertical picture plane is set up such that it is at 45° angles with the vertical walls of the building or interior to display, as shown, for example, in fig. 8.

Fig. 8. Figure 24 from Hans Vredeman de Vries, *Perspective, First book* [1604]

In these cases two distance points on the horizon are the significant vanishing points, while the central vanishing point serves a diminished role. With this special case of two-point perspective, called *diagonal two-point perspective*, a more lively effect can be achieved than with a one-point perspective, while the required knowledge of vanishing points stays the same. Still, only a handful of painters in seventeenth-century Holland tried their hands at this kind of perspective. These were mainly painters of church interiors, for example Gerard Houckgeest (ca. 1600-1661).

This makes it special that two of the preserved perspective boxes showing the interior of a Dutch house each have two standing walls that are painted in diagonal two-point perspective. These are the triangular perspective box in Museum Bredius in The Hague, and the pentagonal perspective box in the Institute of Arts in Detroit.[4] The first dates from about 1670 and is ascribed to Pieter Janssens Elinga (1623 - before 1682), the second from 1663, maybe like the box in London by Samuel van Hoogstraten. The perspective of the box in The Hague will be discussed extensively below. We can be brief about the perspective of the box in Detroit: the front-most, rectangular part is perspective-wise comparable to the rectangular box in Copenhagen, the back part has the same shape as the perspective box in The Hague and the perspective of this part is constructed in the same way.

Angles other than 90° and 45° between the vertical walls of an object to display and the picture plane in Dutch paintings of architecture and interiors are only found in the two perspective boxes with church interiors kept in Copenhagen. The side panels of these triangular boxes are painted in *non-diagonal two-point perspective*. We will revisit this after discussing the perspective box in Museum Bredius.

The perspective box of Museum Bredius

Fig. 9. (Attributed to) Pieter Janssens Elinga, Perspective box with the interior of a Dutch house, circa 1670. Front view. © Museum Bredius, The Hague

The only seventeenth-century perspective box still present in the Netherlands is on permanent display in Museum Bredius in The Hague. The triangular box, which dates from around 1670 and is ascribed to Pieter Janssens Elinga,[5] shows the interior of a rectangular entrance-hall of a Dutch house. The front panel with the peephole in the middle and the light opening above, has not survived.

Fig. 9 shows a front view from great distance. The box is 82 cm high, 84 cm wide and 42 cm deep.[6] The side walls are at a 90° angle with each other and a 45° angle with the front of the box. This made the walls perfect for displaying a rectangular interior in a diagonal two-point perspective, which is actually what Elinga did, even though in his other known paintings he always used one-point perspective.

To understand the perspective of this box we can imagine that the painter worked with a glass box of the same form and shape in a real, existing rectangular interior. It is clear then that the perspective of the painted bottom and ceiling of the box is of the same simple kind as that of the rectangular box of Van Hoogstraten in London discussed above. The legs of the chair partly displayed on the floor have the perpendicular projection of the eye O onto the bottom of the box as their vanishing point. This central vanishing point lies in the middle of the front edge of the bottom of the box.

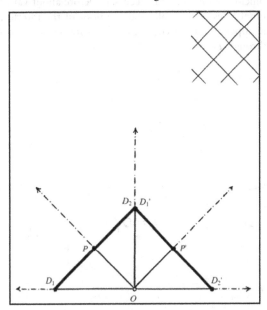

Fig. 10. Top-down view of the imaginary situation of a glass box in the interior displayed in fig. 9, not showing the right proportions. Drawing by the author

More interesting is the perspective on the side walls of the box. Fig. 10 shows a top-down view in which, for clarity, the perspective box and the tiles are drawn bigger in relationship to the size of the interior. The perpendicular projections of the eye onto the side walls have been marked with P and P' and the distance points on the horizon with D_1 and D_2, to the left, and D_1' and D_2' to the right. These points are all in the eye plane, which is the horizontal plane through O. We can see that, because of the unusual shape of the box and the way in which it is positioned in the interior, the lines of sight OP and OP' are parallel to the edges of the diagonally placed floor tiles. From the theorem formulated above then follows that the central vanishing points P and P' are the vanishing points for those edges on the left and right side walls of the box, respectively.

Also, figure 10 shows that the shape of the box and the position of the peephole make that the points D_1 and D_2' lie exactly on the connection between the side walls and the front of the perspective box, while the points D_2 and D_1' coincide with the connection between the side walls. These distance points are more important for the perspective on the side panels than P and P'. After all, OD_1 and OD_2' are not only parallel with diagonals of floor tiles, but also with the horizontal lines on the back wall of the interior. This makes D_1 the vanishing point of these lines on the left panel, and D_2' the vanishing point of these lines on the right panel of the box. Line of sight $OD_2=OD_1'$ is also parallel with diagonals of floor tiles, but moreover this line is parallel with the horizontal lines of the side walls of the interior. So the point $D_2=D_1'$ is the vanishing point of these lines on the side panels of the perspective box.

Fig. 11 shows a cut-out of the sides of the perspective box with a sketch of the principal lines of its painting. The figure also shows the horizon and the vanishing points mentioned above. Use this figure to verify that the construction of Elinga's diagonal two-point perspective coincides with what has been said before about vanishing points of the characteristic horizontal lines of the walls and the floor of the interior. It turns out that the extensions of the upper and lower edge of the left window go through D_1, while those of the left door go through D_2.

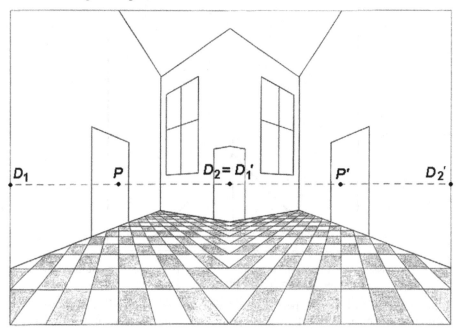

Fig. 11. Cut-out of the sides of the perspective box of Elinga with the principal lines and points of the perspective. Drawing by the author

Fig. 11 can also be used – without having to go to The Hague – to see the effect that the perspective of the side walls of Museum Bredius' box gives when properly observed. To do this, enlarge the figure (preferably to A3 format or equivalent) and print the enlarged figure on thick paper or cardboard. Fold the print along the axis of symmetry. Hold the result open such that both halves, like the side walls of the perspective box, are at a 90° angle with each other. Hold the figure vertically and look at it with one eye level with the drawn horizon, from the point that is at that moment the middle of the connecting line segment between D_1 and D_2'. People with glasses may need to observe from slightly closer or farther. Try until a rectangular interior with a floor of diagonally laid square tiles appears.

The boxes with church interiors

The perspective box mentioned first in this article, with the interior of a Protestant church (see figs. 1, 2) is triangular like the box in Museum Bredius, but has less 'elegant' angles. The side panels are 119 cm high and 75 cm wide, the front panel is 68 cm wide, which means that the side panels are at an angle of 54° with each other and an angle of 63° with the front. Just the two side panels are painted. The unpainted floor panel is invisible as seen from the peephole. The box was created in the period from 1655 to 1660, the painter is unknown. The same goes for the pendant of this box with the interior of a Catholic church, also in the National Museum of Denmark in Copenhagen.

As mentioned, the two triangular perspective boxes in Copenhagen are unique within the seventeenth-century Dutch interior painting arts with regards to the perspective on their side panels. In these boxes, neither the principal vanishing points nor the distance points on the horizon are the vanishing points of the important horizontal lines of the displayed interior. This is evident from fig. 12, where a top-down view sketch is shown of a glass box of the same shape placed in a rectangular area such that the top-down view is symmetric. Just like in the church interiors, we did not place the tiled floor diagonally.

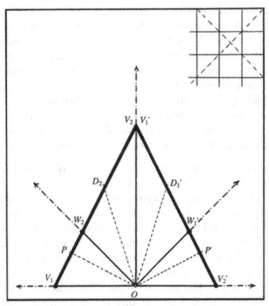

Fig. 12. Top-down view of the imaginary situation of a glass perspective box like those of the National Museum of Denmark in a rectangular interior with square floor tiles. Drawing by the author

In figure 12 we have indicated the principal vanishing point and one of the distance points on both side panels of the perspective box: P and D_2 on the left, and P' and D_1' on the right. Now the lines of sight to these points are not parallel with the edges or diagonals of the square floor tiles or with the horizontal lines on the side walls of the displayed interior and these points therefore do not serve as vanishing points of those lines. We have indicated the vanishing points of the horizontal lines on the back wall of the displayed interior and the edges of the floor tiles parallel to them in the figure with V_1 and V_2', and the vanishing points of the horizontal lines on the side walls of the interior and the other edges of the tiles with $V_2 = V_1'$. Note that these vanishing points again lie on the edges of the panels of the perspective box. We call the vanishing points of the diagonals of the floor tiles W_2 and W_1'. Fig. 12 shows that these points are *not* the midpoints of the line segments $V_1 V_2$ and $V_1' V_2'$. By the way, the unknown painter of the Copenhagen perspective boxes would not have concerned himself with the placement of W_2 and W_1'. He does not seem to have attempted to construct the tile floor exactly (see fig. 2).

Modern perspective boxes

Towards the end of the seventeenth century, the general public appeared to have lost interest in perspective boxes. After that, they were no longer made in that particular form. In current Dutch secondary education they play a renewed part, because perspective has become an optional subject in some mathematics courses.[7] With this subject the construction of a perspective box turns out to be a popular final project that gives nice results.

The students in question are inspired not only by seventeenth-century perspective boxes. Also folding cardboard perspective boxes with the interior of a bakery that were around 1980 in the Netherlands only locally and briefly in use as pastry boxes (fig. 13), serve as motivating examples.[8] These boxes had a peephole, which made them function the same as the seventeenth-century perspective boxes. When figure 13 is enlarged so that it is 34 cm high and 26 cm wide and printed on thick paper or thin cardboard, a *baker's box* to scale can be obtained with a bit of cutting (don't forget the slits of the peephole left and right) and folding. When observed through the peephole, the baker's legs are no longer strangely long and the cupboard up against the wall on the right side is just rectangular. Drawing such a rectangular cupboard, spread across three side panels, is a good exercise before students make their own perspective box.

Students usually use a shoe box without the lid as the basis for such a perspective box. They make a peephole in one of the small vertical walls. This way the perspective is comparable to that of the rectangular perspective box in Copenhagen and the baker's box. The students replace the old-Dutch interior of a house or the bakery with, for example, displays of their own bedrooms or their school. Some shoe boxes can be folded and unfolded multiple times without any trouble. This is useful while drawing, checking, erasing and re-drawing. Those who want to attempt the challenge of making a triangular box with a right angle like the seventeenth-century perspective box in The Hague or maybe even with acute angles like the two perspective boxes with church interiors in Copenhagen, have to first create a box in that shape. In these cases it is even more important that the box be easy to fold and unfold. Mistakes while applying the construction rules are more easily made than with rectangular boxes, so checking often through the peephole is even more recommended.

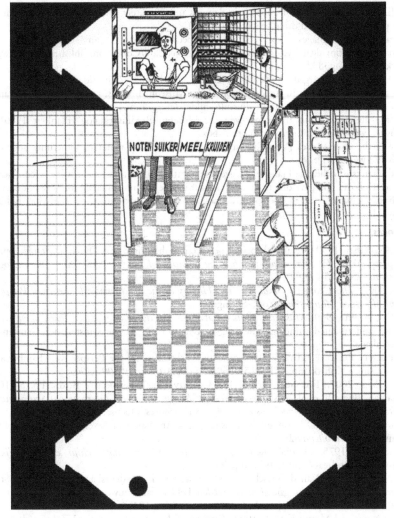

Fig. 13. Cut-out of the baker's box

That students sometimes do not construct the hardest parts of their perspective boxes, but rather just wing it while they are looking through the peephole, should be forgiven. Doing that they follow in the footsteps of the seventeenth-century perspective painters and the unknown twentieth-century artist of the baker's box.

Acknowledgments

This article is an edit of [Verweij 2001], translated by Lodewijk Vöge, son of the author. With thanks to Claus Jensen, Nibe, Denmark, who suggested some improvements in content.

Notes

1. More data and images of surviving perspective boxes than this article contains can be found in [Andersen 2007], [Blankert 1980], [De la Fuente Pedersen 2005], [Jensen 2007], [Koslow 1967] and [Leeman 1975]. Only [Andersen 2007] and [Jensen 2007] also have a thorough analysis of the perspective in the discussed box, which is in both cases the perspective box made by Samuel van Hoogstraten.

2. In [De La Fuente Pedersen 2005] it is claimed, unfortunately without a solid argument, that this perspective box had three peepholes (p. 158).
3. Data and images of this perspective box can be found at: http://www.nationalgallery.org.uk/cgi-bin/WebObjects.dll/CollectionPublisher.woa/wa/work?workNumber=ng3832.
4. See http://www.dia.org/the_collection/overview/viewobject.asp?objectid=48296.
5. For the works of Elinga in the collection of the Museum Bredius, the perspective box among them, see: http://www.museumbredius.nl/schilders/elinga.htm.
6. All the literature about this box, even the Museum Bredius catalog [Blankert 1980] and website, swap the height and width of the box.
7. One of the courses uses [Verweij and Kindt 1999] as study material, sometimes augmented with [Verweij 2001].
8. With thanks to colleague Hans ter Heege, Freudenthal Institute, Utrecht, who saved a number of baker's boxes and gave one to the author of this article. Unfortunately we have not been able to find out who made the perspective drawings for this box and who, if anyone, holds the copyright.

References

ANDERSEN, Kirsti. 1990. Stevin's theory of perspective: the origin of a Dutch academic approach to perspective. *Tractrix, Yearbook for the History of Science, Medicine, Technology & Mathematics* 2: 25-62.
————. 2007. *The Geometry of an Art. The History of the Mathematical Theory of Perspective from Alberti to Monge.* New York: Springer.
BLANKERT, Albert. 1980. *Museum Bredius, Catalogus van de schilderijen en tekeningen*, second, revised and updated printing. The Hague: Dienst voor Schone kunsten der Gemeente 's-Gravenhage.
DE LA FUENTE PEDERSEN, Eva. 2005. Cornelius Gijsbrechts and the Perspective Chamber at the Royal Danish Kunstkammer. Pp. 152-160 in *SMK Art Journal 2003-2004.*
JENSEN, Claus. 2007. The Geometry of 17th Century Dutch Perspective Boxes. Pp. 89-106 in Sriraman, Bharath, et al., eds., *Proceedings of Macas 2, Second International Symposium on Mathematics and its Connections to the Arts and Sciences*, Odense, Denmark, 2008.
KOSLOW, Susan. 1967. De wonderlijke Perspectyfkas: An Aspect of Seventeenth Century Dutch Painting. *Oud Holland* 82: 35-56.
LEEMAN, Fred. 1975. *Anamorfosen, Een spel met waarneming, schijn en werkelijkheid.* Amsterdam: Andreas Landshoff Production.
STEVIN, Simon. 1605. Van de verschaeuwing, Eerste bouck der deursichtighe. In Simon Stevin, *Derde stuck der Wisconstighe ghedachtenissen.* Leiden: Jan Bouwensz.
VERWEIJ, A. 2001. Perspectief in een kastje. *Nieuwe Wiskrant* 21, 2: 6-16.
VERWEIJ, Agnes and Martin KINDT. 1999. *Perspectief, hoe moet je dat zien?* Zebra-reeks volume 2. Utrecht: Epsilon Uitgaven.
VREDEMAN DE VRIES, Hans. 1604-1605. *Perspective deel I en deel II.* Leiden: Hondius, Henricus. (Rpt. 1979 in facsimile edition with explanation by Peter Karstkarel. Mijdrecht: Tableau BV).

About the author

Agnes Verweij holds a M.S. degree in Mathematics from Leiden University. She is a senior lecturer of mathematics and mathematics education at both the Department of Applied Mathematics and the Department of Science Education of Delft University of Technology in The Netherlands. Her research interests and publications cover such different topics as the use of computer algebra in secondary and higher mathematics education and the combination of geometry education and aspects of art in secondary schools. She is the leading author of the book *Perspectief, hoe moet je dat zien?* (co-author Martin Kindt, Epsilon Uitgaven, Utrecht, 1999) used for teaching and learning perspective as an optional topic in the higher levels of Dutch secondary schools. Lately she collaborated as the expert on geometry, and in particular on perspective, in a team that designed educational material for a new level of mathematics in Dutch secondary schools for prospective students in cultural and social sciences.

Open Access This article is distributed under the terms of the Creative Commons Attribution Noncommercial License which permits any noncommercial use, distribution, and reproduction in any medium, provided the original author(s) and source are credited.

Tessa Morrison

The School of Architecture
and Built Environment,
The University of Newcastle
Callaghan, NSW, 2308
AUSTRALIA
Tessa.Morrison@newcastle.
edu.au

Keywords: Juan Bautista
Villalpando, architectural
drawing, seventeenth
century, perspective, optics,
geometry, perspective
theory

Research

Juan Bautista Villalpando and the Nature and Science of Architectural Drawing

Abstract. In 1604, Jesuit priest and architect Juan Bautista Villalpando published *In Ezechielem Explanationes*, a massive three-volume scriptural exegesis on the Book of Ezekiel. Volume Two was dedicated to the reconstruction of Ezekiel's vision of the Temple of Solomon and consists of five books: on the prophecy on the Temple; its plan and reconstruction; the justification of the reconstruction; its bronzes and ornamentation, and an entire book on the nature and science of architectural drawing. Initially the latter appears out of place in a Scriptural exegesis but he explained that the purpose of this book was to provide a guide for theologians so that they can form a mental idea or image of Temple, for their understanding and enlightenment of the entire Temple. However, throughout the text he points to the utility of the book to architects. For Villalpando the laws of optics were essential to the norms of perspective. Moreover, the sense and structure of seeing was a crucial element to the norms of mathematic and architecture, it is also a central theme in his theology. This paper examines his theory and his proposal of perspective for architecture drawing.

Introduction

In Ezechielem Explanationes was to be a collaboration by two Spanish Jesuit priests, Jerónimo del Prado and Juan Bautista Villalpando. It was Prado's project and Villalpando's role was a minor one: he was to provide a reconstruction of the Temple to coincide with Prado's text. However the early death of Prado, after the completion of Volume One, left the entire project for Villalpando to complete on his own. Volume Two and Three are entirely written by Villalpando [Morrison 2009: 1, 12-13]. Surviving correspondence reveals that Prado and Villalpando disagreed on the design of the Temple [Arroyo-Furphy and Tolton 2009]. It is very possible that if Prado had lived Solomon's Temple would not have dominated the entire Volume Two and the reconstruction would have been subordinate to Prado's Scriptural exegesis.

In Ezekiel's vision of the Temple his spirit is guided by an Angel through the Temple, measuring the walls, the courtyards, the altar and so on. The prophet does not claim that the temple he described is Solomon's, he only claimed that it is a vision of the Temple of Jerusalem. Traditionally the description in the Biblical texts of 1 Kings and the Second Book of Paralipomenon (Chronicles) are taken to be Solomon's Temple and the text of Ezekiel to be a vision of a future Temple. However, Villalpando claimed that Ezekiel's vision was the Temple of Solomon and he made no distinction between Solomon's Temple and the vision of Ezekiel. He used 1 Kings, Second Book of Paralipomenon, and the Book of Ezekiel, as describing the same building [Villalpando 2009: 133-141]. When the measurements of these Biblical texts did not equate to each other, Villalpando made complex explanations to demonstrate that if they were examined in a different way, they were in fact the same [Villalpando 2009: 349-361]. Villalpando's use of Ezekiel's vision caused an outcry within the Jesuit order. Pope Sixtus V set up an Inquisition

Nexus Network Journal 12 (2010) 63–73   NEXUS NETWORK JOURNAL – VOL.12, No. 1, 2010 **63**
DOI 10.1007/s00004-010-0017-5; *published online* 6 February 2010
© 2011 Kim Williams Books, Turin

commission, headed by Cardinal Toledo, to examine Villalpando's thesis. At the same time General Father Aquaviva also conducted an inquiry by appointing a committee of the Father of the Society of Jesuits to examine Villalpando's orthodoxy. In the end Villalpando was cleared of any suspicion of heresy, but he continued not to make any distinction between Solomon's Temple and Ezekiel's vision [Morrison 2009. 10-11].

The illustrations are an integral part of the entire text of *Ezechielem Explanationes*. Villalpando claimed that:

> I have not studied it (the Temple) in order to re-establish the old glory of the Temple, but in order to interpret the texts of the Scriptures that contain the sublime mysteries of our religion; my intention was to clarify everything that is the object of our sensorial comprehension of this information to discover other more divine elements [Villalpando and Prado 1604: II, prologue] (n.b. All quotations from *In Ezechielem Explanationes* are translated from Latin by the author).

To "clarify everything that is the object of our sensorial comprehension" there are forty-eight copperplate engravings, twenty of these folded out with some that are over one and half metres wide. These engravings are of exceptionally fine quality, and included the plans, elevations and perspective drawings. Villalpando gathered together a team of highly skilled artists and engravers from Rome and Flanders at the workshops in some rooms of the Roman College, his residence in Rome. None of the engravings are signed and in his text Villalpando claimed "I have drawn", clearly claiming to be the artist. Although Villalpando closely supervised the works he did not do any of the etching himself [Morrison 2009: 18-20]. One of the engraved plates, "The Egyptian Night," was inscribed "Graphic method f," the name scraped and burnished away with great effort, although the deliberate deletion is partially visible [Ripoll 1991: 270]. Perhaps Villalpando felt that it was "his" reconstruction and that he was the artist of this reconstruction. However, Villalpando believed that the original plans of the Temple were drawn by the Divine hand of God:

> It deserves also our praise that God with his hand drew the design, the figures, the location of all the elements, the graphic plan, the elevations and perspectives; and besides that God described all of it with abundant comments that were delivered to David, at the same time, were given to Solomon so that the artisans could carried it out all to perfection [Villalpando and Prado 1604: 104].

For Villalpando it was the "Architecture of Theology" [Villalpando and Prado 1604: II, Prologue] which revealed the mind and plans of God. To contemplate or meditate to visualize, was an attempt to reveal the truths within. Ignatius of Loyola, founder of the Jesuit order, promulgated this concept in his *Spiritual Exercises* of 1544-5, a set of contemplation exercises which emphasise the use of the five senses of the imagination. Seeing is the dominate sense: his contemplation exercises begin with the words, "Composition, seeing the place." Ignatius stated that, "composition consists in seeing through the gaze of the imagination the material place" [Loyola 1996: 294]. On a meditation of hell he claimed that the "composition here is to see with the eyes of the imagination the length, breadth and depth of hell" [Loyola 1996: 298]. In a prayer of the senses he asserted that the aim was:

To see the persons with the imaginative sense of sight, meditating and contemplating their circumstances in detail, and to draw some profit from what I see [Loyola 1996: 307].

Ignatius drew "some profit" not only from "seeing" the narrative but also places and buildings. To see or conceive these holy places or narratives was to come closer to an understanding of the Sacred Scriptures.

Villalpando perceived the significance of the standards and norms of the Architecture of Theology, which would clarify the conditions and the essential requirements of the Architect. He formed a very concise theory on the nature and science of architectural drawings and its importance. This theory was "to instruct a theologian and not a labourer..." [Villalpando and Prado 1604: II, 47], and the images were "in order to guide the theologians so that they can form a mental idea or image of Temple..." [Villalpando and Prado 1604: II, prologue], for their understanding and enlightenment of the entire Temple; but although Villalpando emphasised the spiritual application of his theory the practical application of it was ever present. The sacred was the model for the profane; and architectural drawings were an essential form of communication of the architect. The architect must not only "see" by contemplation the entire building to be built, but must also be able to communicate that vision.

Villalpando on the nature and science of architectural drawing

Villalpando considered Vitruvius's nature of architectural drawing, called in Greek ἰδέας "Ideas," to be "the description of the plan of a building (*Ichnographia*), the elevation (*Orthographia*) the perspective (*Scenographia*)" [Villalpando and Prado 1604: II, prologue]. He quoted Vitruvius's description of perspective: "It is the drawing of the facade and of the side parts of the building as being moved away toward a point where all the visual lines corresponds" [Villalpando and Prado 1604: II, 62]. Villalpando claimed that interpreters of Vitruvius, including Daniele Barbaro, believed that the term should be *sciographia*, and not *scenographia*. Sciographia is a "term that signifies a sketch, a planned and shaded painting, where the lines represent an empty and deceitful form of the objects" [Villalpando and Prado 1604: II, 62], or what Cicero calls a "shaded sketch." But this form of drawing is only useful to painters and not to architects. It is only

> ...perspective that puts in sight the whole and all of the parts of the building, its limits, its projections, the doors and the gates as well as all the other elements, by means of its graphic expression of lines and although all these lines concur in a point; on the other hand, in a painting it is simply shaped so that you can barely distinguish the limits that are moving away from the objects and the ones that are in sharp view [Villalpando and Prado 1604: II, 62].

Villalpando believed that drawings that were guided by the rules of optics were the only useful tool for producing architectural plans. Painting can give a lifelike image but it was not a good tool for the architect, and although painters and architects can have the use of perspective in common there is a distinct difference in its use. The drawings of an architect results in a building that is a copy or perfect imitation of the drawing produced, while the drawings of the painter consist in an imitation of the building, which already exists, reproducing its colours, the stone and the textures [Villalpando and Prado 1604: II, 42]. In this discussion of perspective for painting and architecture, Villalpando appears to be criticising Alberti for not having drawn a distinction between the use of perspective and architecture in *De Pictura* [Alberti 1972]; however, Villalpando does not mention Alberti by name.

But light and the science of vision are the study of nature and therefore it is considered to be a study of philosophy. The philosopher considers the nature of light, its movement, its production or its corruption of objects, whereas the "teacher of perspective examines the vision or light, its limits, its lines, its surfaces, its points, its figure, its sections, its directions in straight lines, and he supports the nature of light as the Philosophers say" [Villalpando and Prado 1604: II, 48]. There are two categories of science of light: the first is to speculate, this is the realm of the philosopher; the second is to make it pragmatic, this is the realm of the architect.

Villalpando used Aristotelian terminology, such as "species" and "power" [Aristotle 1965], in a framework of Euclidian optics. "The visual thing is called power, the visible thing is called object, because it acts on the power through its form" [Villalpando and Prado 1604: II, 52]. The 'forms' or 'species' are an immaterial consequence of the object that illuminates the environment. These 'forms' are what the object projects into the environment. This projection of the 'forms' or 'species' into the environment stimulates the 'power' so that the vision can take place. To Villalpando the 'forms' or 'species' and the 'power' are two processes that make sight possible; one projects the form of the object the other makes that projection visible and it would not be possible to see the object if the species of the object was not transmitted to the power. Villalpando does not fully explain the 'power' but leaves it as a mystical or divine process which is essential to sight.

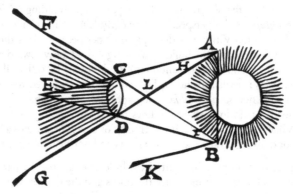

Fig. 1. A graphic demonstration of the angles of movement of light [Villalpando and Prado, 1604: II, 50]

Villalpando engages in a lengthy discussion on the angles of the movement and action of light, and more importantly the angles of the areas in total shadows or defused light (fig. 1). This theory of physics – that light does not mix – is shared by the teachers of perspective because:

> It is observed that light that moved from the two luminous bodies follow a dual pathway, each in a straight line on the paths to each one of the luminous bodies and also because the shadows are moving separately from the opaque bodies [Villalpando and Prado 1604: II, 50].

The shadows will always be larger than the body that produces them when the luminous body is smaller than the body that produces the shadow. In fig. 1, if the only light comes from the arc AB, it will project the shadow of body CD within the space

FCDG, which will become increasingly larger the further it moves away from the straight line CD. This is shown by the fact that straight lines EA and GH are extended and intersect at point A, so that they form a triangle whose base is CD; considering the shadow behind this base, it can be seen that it is impossible to draw a line equal to or smaller than CD that touches both lines CE and DG, as can be proven by lines parallel to CD.

The species are projected into the environment, as light is, in a straight line:

> The species of the objects are transmitted only to those places in which the objects themselves are present, that is to say, by means of a few straight lines from the objects towards the places to where the species must be transmitted [Villalpando and Prado 1604: II, 49].

Vision is carried out in the same way:

> ...around the straight line that would be drawn from the object seen to the centre of the visual power through the pupil of the eye: this is what the teachers in perspective call 'perpendicular line' to this straight line. It should be understood that through this line the visual species arrives to the eye and also through this same line leave the 'spirits' positioned from the eye to the object seen [Villalpando and Prado 1604: II, 50].

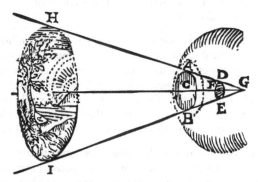

Fig. 2. Villalpando's cone of vision [Villalpando and Prado, 1604: II, 57]

Fig. 2 shows Villalpando's examinations of the vision of objects. He demonstrated that the laws of optics affirm that the rays are emanated from the eye by means of lines, separated from each other by a small distance. Those rays touch the object seen in the cone: the base of this cone is the object seen and the vertex is situated in the eye. The greater the angle of this vertex, the greater the area seen:

> The circle AB, is the hole of the pupil, whose centre will be C; and the small circle DE, is the liquid organ of the vision, the centre will be F; CG [n.b. This was written as CE in the original text] is joined by a straight unlimited line that is called axis or perpendicular line, and by the points A and D of the circles AB, DE, will correspond with the axis; they will be joined through a straight unlimited line AD that will cut to the axis in the point D. And remaining motionless at the point G surrounding itself the straight one GA until it returns to its place of origin: there remains to describe, a cone, by the movement of this line and this cone covers

everything that the eye can see in a straight vision; if we think that the cone finishes in the visible object, the base of this cone will be HI [Villalpando and Prado 1604: 57-58].

According to Villalpando, the architect cannot design a building of dignity and elegance if he does not have in his mind the totality of the building he wants to construct.

But the architect must translate their vision from a mere consideration of the distinct parts which does not seem sufficient to contemplate mentally the symmetry, the beauty, the order, the proportion and other aspects that are given in the confrontation of all the building and of its parts. The architect must convert their mental image by means of lines so that once captured in a drawing one could see such images and you can see another and be able to compare them, judge and perfect them [Villalpando and Prado 1604: II, 61].

Such drawings could be constructed in perspective:

Perspective can be understood to be like the description of the stage. The stage is that part of the theatre that extends between the two extremes, that is to say, all that appears before the eyes of the one that contemplates that middle part of the theatre, and this can neither take shape nor be expressed with either lines or with colours unless the method of optics is realized [Villalpando and Prado 1604: II, 63].

The architect must translate his ideas into perspective, which adheres to the norms of optics, that is, to provide the division or section of the cone of vision and of the proposed plan. The purpose of perspective is to place the lines on a given surface, "so that its species excite the vision exactly the same as it excited the species of the object proposed" [Villalpando and Prado 1604: II, 63]. The eye contemplates exactly the same thing when it examines the lines planned in a given surface that includes the object itself. It must preserve the same order and the same distribution that are preserved in the objects:

Finally we have concluded, and this should remain perfectly clear that you will not see the distances of objects and that species normally excite the vision in this way, since they occurring away from an object, an object that is close, although they excite the vision with greater force than the distant objects more closely represent the parts of the object, which they cannot achieve [Villalpando and Prado 1604: II, 63].

Villalpando claimed that this can be demonstrated by means of mathematics; the straight lines are laid out from all the parts of the object toward the eye, where its species are received; in considering any surface between the object and the view sighted, this will cut all the lines cited and they will reach the eye because the lines have been planned from the object across the surface. They will preserve the same disposition and distance for what it refers to as the limits of the lines; limits that have been received in the eye turns out to be totally extrinsic, although there is placed some prolonged lines, or equal ones, or some shorter and others even more prolonged:

It follows that if someone marks on the surface the limits of the lines, so that they can excite the vision, they will produce the same vision that were obtained earlier when they were coming from the object. But if a few lines are joined at the said limits of these lines, not only would they represent the indivisible limits of the visible object, but also the limits of their length: we call these lines and the physical points visible, not mathematical. These designed lines according to what we have just said, if they are distinguished also by their shadows, in such a way they would deceive the human eye with objects similar to the one that remain represented by the lines, they would seem authentic [Villalpando and Prado 1604: II, 63].

Vitruvius stated that the architect should know the art of painting in order to plan with facility. Villalpando claimed that this appears to imply drawing as a special class of instruction which is necessary for carrying out the works of the plan, elevation and perspective. To Villalpando it appears that Vitruvius evidently:

...separated the skill of the science of optics, from the sciences of the architect, he specified them respectively by separating optics and drawing, granting to drawing the different sketches of the works and to optics the study of the brightness in the buildings and in other areas [Villalpando and Prado 1604: II, 63].

Villalpando points out that this does appear to be the opposite to what he is saying since the purpose of his book was to recognize the assumptions that optics would be able to explain the descriptions of architectural drawing. He continued,

...but if this theme is examined with attention, it will be discovered that there is something in common between optics and drawing, as between that part of geometry that deals with the intellectual refinement and the part that its purpose is to execute the work; to this part are called the practical part, and to the other speculative part [Villalpando and Prado 1604: II, 63].

Although geometry is called the practical part, it cannot be understood completely if it does not adhere to the principles of speculative geometry. On the other hand it cannot be understood as some form of drawing that is the practical part of optics without some principles which have been taken from speculative optics.

Perspective represents the image of the building in totality, already constructed, with its façade and its sides in a single image:

The presence of this building remains included in the cone (of vision) and if you think therefore it is cut by a flat surface, for example by a board or a parchment or a papyrus, and if you can image that the lines of all the limits of the building leave its track or plan in the surface, these tracks represent the same building before the eyes placed in the same distance, in the same way that this building remained represented by its own species. The figure on this board will be called perspective, we already know how it differs from a painting: a painting realizes the same parts of the building with its colours and textures, for example the stones, the wood, the grids of iron; on the other hand the perspective represents the distinct parts of a building by means of lines and shadows [Villalpando and Prado 1604: II, 64].

Fig. 3. The view of the vault, of the walls, of the pavement of the sacred Sanctuary and Testament together with the Cherubim [Villalpando and Prado, 1604: II, no page number]. Reproduced with permission from the British Library

For Villalpando, perspective was not restricted to the facades and the sides but also included the pavement and the ceiling, but "Vitruvius does not mention this because he appeared to consider that this was the function of the plan" [Villalpando and Prado 1604: II, 65].

However, this form of perspective requires a great deal of skill and art. Villalpando's purpose was to present a perspective of the entire Temple of Solomon. He claimed that:

> Frequently I have endeavoured in proposing an single perspective of the entire building and I have obtained it, but times also I desisted from this

pledge, defeated by the difficulty of such a perspective and by the great deal of time that it requires of me; although I have not been able to translate the perspective of the entire temple, I have completed a perspective of the noblest and most important part of the Temple, that of the Sacred Sanctuary entitled: view of the vault, of the walls, of the pavement of the sacred Sanctuary and Testament together with the Cherubims [Villalpando and Prado 1604: II, 65].

Although Villalpando stated that he had not lost hope of drawing a perspective of the entire Temple, the image of the Sacred Sanctuary is the only perspective in the entire three volumes of *In Ezechielem Explanationes* (fig. 3). It is a haunting and mysterious interior of the most sacred part of the temple; it contains the Ark of the Covenant and the Cherubim that guard it. The Ark is not of the Biblical dimensions and Cherubim are more human and serene than the four-faced guardians described in the Bible, which were part lion, part man, with the wings of an eagle and the claws of a lion [Villalpando and Prado 1604: II, 88], in many ways this engraving contradicts the text that it is illustrating.

But Villalpando was attempting to propagate an entire theory of the nature and science of architectural drawing and not just perspective itself. The plan, elevation and perspective were all forms of optics and thus are all forms of perspective.

However, he defined two forms of vision. One is common and natural, in which sight is less than the object seen. The other, which can be imagined by using the example of light when the eye is supposed to be equal to the object seen: the visual species remains in the object but a cylinder will appear to be a circle or a rectangle. He referred to this second form of vision as "hypothetical vision." The perspective requires the first form of vision; the plan and the elevation require the second:

However, in this section of the cylinder itself is something that is unique and exclusive to the plan and the elevation, and therefore we understand that it always takes place in a plane formed at right angles to the parallel figure; the contrary thing happens, as we have said, in the perspective, because there we assume that the cone of vision is cut by a plane formed by any and every kind of angles. From this it is deduced that everything that is parallel to the vision is formed exactly equal, in the elevation and in the plan, to the same elements or parts of the building and keeping between itself the same proportion, therefore is originated from the cut or section, such cuts of a parallel figure is what we have referred to as the plan and the elevation [Villalpando and Prado 1604: II, 66].

Villalpando goes to great length to explain this "hypothetical vision," using the same Aristotelian terminology that he used in his theory of perspective. He continually referred to it as "vision," only using the term "hypothetical" once, despite the fact that it has no focal point, all lines or rays are parallel, and it in no way relates to the vision he previously described. Although he refrained from using the term "cone of vision" his repeated use of "vision" implies some relation with perspective. Villalpando's use of Aristotelian terminology and his convoluted explanations often makes very simple concepts seem extremely complex and confusing. But he ends his description of the nature and science of architectural drawing with a clear example of an elevation (fig. 4):

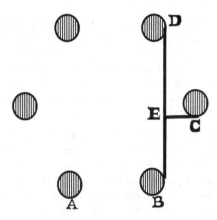

Fig. 4. Example of hypothetical vision [Villalpando and Prado, 1604: II, 67]

It [the elevation] can be seen with clarity with the following example: ABCD be the building, of six columns A, B, C, D, etc, separated by equal inter-columns. If its facade is observed AB, since the vision is carried out through parallel lines cut at right angles, there will be seen the inter-column AB, which is equal to the inter-column BC but it will not appear to be the same without depending on the distance of CE that lays parallel to AB; BD is in the line of the vision and perpendicular to CE next to the one that the section is carried out. As one cannot draw any perpendiculars to the parallel lines that are greater than the distance of the columns, they are considered almost perpendicular, it turns out that we cannot see any inter-column greater that the one that will in fact be built in the building [Villalpando and Prado 1604: II, 67].

Conclusion

Throughout the text of Volume Two of *In Ezechielem Explanationes*, Villalpando quoted copiously from other authors, either to support his arguments or to refute theirs. Yet in Book Two there is a stark contrast in the lack of authors quoted. Villalpando was aware of the theories of Alberti and Serlio, both are mentioned elsewhere, but he does not mention either author in his discussion of perspective. Although he claimed that his purpose in Book Two was to create an "Architecture of Theology," so that through understanding the planning of Solomon's Temple the theologian could understand the Scripture, primarily Book Two is addressed to architects and involves the practical application of perspective.

For Villalpando perspective was the most adequate form of architectural drawing it does justice to the symmetry but "the perspective cannot achieve the adequate proportion of all the work and of all their parts" [Villalpando and Prado, 1604: II, 67]. All three forms of architectural drawing are required. The perspective drawing was required for the first form of vision – the cone of vision; while the plan and the elevation are required for the second form of vision – the cut section of the cone of vision, which in the case of the plan and elevation is a cylinder.

For Villalpando the profane is a copy of the sacred. To understand the practical science, the architect of the copy should to be able to read and comprehend architectural drawings as if they were the sacred. After all, the

...figure or image of the house that was presented by Ezekiel, and that we want to reproduce according to our possibilities, had to be a figure or design to a smaller scale. That is to say, reduced to a smaller form; unless, by means of this term, that is to be understood in the same way that Vitruvius indicates, when he speaks of the 'reduced plan,' or of reducing the elevation of a building [Villalpando and Prado 1604: II, 91].

Ezekiel reduced the vision of the Temple to a reduced plan to convey the sacred by only using text. The "composition" of the Temple was seen through the images of *Ezechielem Explanationes,* which were a mixture of plans, elevations and a single perspective which made it possible to contemplate the temple and guide people through the text of Ezekiel. Villalpando perceived perspective as a Divine invention from God which was inscribed into the sacred text of Ezekiel, which was a guide for architects to use in their profane architecture and in their attempt to mimic the divine.

References

ARISTOTLE. 1965. *De Anima.* R. D. Hicks trans. Amsterdam: A. M Hakkert.

ARROYO-FURPHY, Susana, and Laura TOLTON. 2009. Appendix A: Documents Pertaining to the Dispute between Father Villalpando and Father Prado. Pp. 573-78 in *Juan Bautista Villalpando's Ezechielem Explanationes: A Sixteenth-Century Architectural Text.* Tessa Morrison, ed. Lampeter, New York: Edwin Mellen.

LOYOLA, Saint Ignatius of. 1996. *Personal Writings: Reminiscences, Spiritual Diary, Selected Letters Including the Text of the Spiritual Exercises.* Joseph A Munitiz and Philip Endean, trans. London: Penguin.

MORRISON, Tessa. 2009. *Juan Bautista Villalpando's Ezechielem Explanationes: A Sixteenth-Century Architectural Text.* Lampeter, New York: Edwin Mellen.

RIPOLL, Antonio Martinez. 1991. El 'Taller' De Villalpando. Pp. 243-84 in *Dios Arquitecto.* Juan Antonio Ranirez, Rene Taylor, Andre Corboz, Robert Jan Van Pelt and Antonio Martinez Ripoll, eds. Madrid: Ediciones Siruela.

VILLALPANDO, Juan Bautista, and Jerónimo del PRADO. 1604. *Ezechielem Explanationes Et Apparatus Urbis Hierolymitani Commentariis Et Imaginibus Illustratus.* 3 vols. Rome.

About the Author

Dr Tessa Morrison is an Australian Research Council post-doctoral fellow in the School of Architecture and Built Environment at the University of Newcastle, Australia. Her academic background is in history, mathematics and philosophy. Her current research project focuses on sixteenth and seventeenth century sacred architecture. She has published articles on geometric and spatial symbolism, and architectural history; her recent book publications are *Juan Bautista Villalpando's Ezechielem Explanationes: A Sixteenth-Century Architectural Text* (Lampeter, New York: Edwin Mellen, 2009) and *Labyrinthine Symbols in Western Culture: An Exploration of their History, Philosophy and Iconography* (Saarbruecken: VDM Verlag, 2009).

José Calvo-López

Escuela de Arquitectura e
Ingeniería de la Edificación
Universidad Politécnica de
Cartagena
Ps. Alfonso XIII, 52
30203 Cartagena SPAIN
jose.calvo@upct.es

Miguel Ángel
Alonso-Rodríguez

E. T.S. de Arquitectura
Technical University of Madrid
Avda. Juan de Herrera, 4
28040 Madrid SPAIN
miguel.alonso@upm.es

Keywords: Paolo Uccello; Piero
della Francesca; Daniel Barbaro;
Jacopo Torni l'Indaco; Pedro
Machuca; Luis Machuca;
Junterón Chapel in the Cathedral
of Murcia; Palace of Charles V in
Granada; perspective; descriptive
geometry; torus; surfaces of
revolution

Research

Perspective versus Stereotomy: From Quattrocento Polyhedral Rings to Sixteenth-Century Spanish Torus Vaults

Abstract. Quattrocento perspective and Spanish sixteenth-century stereotomy share a number of concepts, problems and methods, although there seems to be no direct substantial connection between them. This suggests the existence of a common source, but it is not easy to identify it. Neither classical geometry nor the mediaeval practical geometry tradition include a word about orthographic projections, rotations or projection planes. Thus, mediaeval construction shop practices furnish the most probable common source for perspectival and stereotomic methods. Curiously, these practices are seldom mentioned in the exhaustive literature on perspective; even the use of orthogonal projection, although impossible to deny, is not often stressed. On the other side, Gothic tradition is recognised, at least in Spain, as an important source of Renaissance stereotomic methods. By contrast, the role of perspective and Italian and Italianate artists, which has been downplayed so far, should also be taken into account as a source of Renaissance stereotomy.

Introduction

When trying to assess the sources of the geometrical knowledge of Renaissance stonemasons in his now classic book *L'Architecture a la française*, Jean-Marie Pérouse de Montclos made a startling suggestion: taking into account the lack of a clear description of orthographic projections in Mathes Roriczer's booklets, stonecutters could have taken their geometrical expertise, at least in part, from the scientific description of space in perspective treatises, ranging from Euclid's *Optics* to Niceron's *Perspective curieuse* [de Montclos 2001: 184-185].

At first sight, the idea seems to be rather far-fetched. The central problem in classic masonry construction – the division of an architectural member into voussoirs and the geometrical control of the dressing process by means of templates and other methods, known from the seventeenth-century on as stereotomy – is always solved by means of double or multiple orthographic projections (see for instance [Palacios 2003; Sakarovitch 1997; Rabasa 2000]). The stonemason starts the construction process of a complex arch or vault by preparing a full-size tracing involving at least two orthographic projections of the member, a plan and an elevation or a cross-section. At this point, the mason can choose between two basic methods. When using the direct method (*taille par panneaux, labra por plantas*), the stonemason must prepare full-scale templates of the voussoir faces, either by rotation or development of the distorted faces of the voussoir. Alternatively, the mason can use the squaring method (*dérobement, équarrissement, labra por robos*). In this method, orthographic projections are essential, since the mason materializes the projection planes of the lines in the tracing by means of the square, reversing the projection process, in order to dress the voussoirs [Palacios 2003: 18-21; Calvo 2003].

In any case, both the direct and the squaring methods start from orthographic projections, that is, from a particular case of parallel projections. Of course, this is not the case in linear perspective, since all projecting lines converge in the station point or centre of projection, in the same way that the generatrices of a cone converge in its vertex. To put it into Albertian terms, all the rays that form the visual pyramid converge in the eye of the observer. Thus, linear perspective belongs to central or conical projection, as opposed to parallel or cylindrical projection.[1]

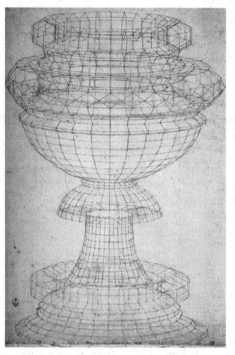

Fig. 1. Drawing of a chalice. Uffizi 1758 A

However, there is at least a connecting point between Renaissance perspective and stereotomy. Piero della Francesca [c. 1480: 37v-41 v], takes great pains to explain the construction of an unusual figure, a wooden polyhedral ring (*mazzocchio, torculo*), whose only function seems to have been to support the clothes of some elaborate headgear of the period (see figs. 7, 8, 9 below). A number of perspectival studies for rings in the Uffizi and the Louvre, as well as *Chalice* surrounded by three rings, also in the Uffizi suggest that this figure acted as a typical Renaissance perspective exercise (fig. 1; see also figs. 5, 6 below). These rings also appear in some *intarsia* or inlaid wooden panels of the period, in particular in the *studioli* or private cabinets of Federico de Montefeltro at Urbino and Gubbio, playing the role of a symbol of the geometry amongst the liberal arts. They are also to be found around the necks or over the heads of some characters in Paolo Ucello's paintings, such as the *Flood* in the Chiostro Verde in Santa Maria Novella (see fig. 10 below) or the *Battle of San Romano* (see also [Davis 1980; Kemp 1990: 32-33, 44; Evans 1995: 173-175; Raggio 1996; Davis 2001; Roccasecca 1998; Roccasecca

2000; Roccasecca 2001a; Roccasecca 2001b; Roccasecca 2001c; Field 2005: 168; Esteve 2007]).

Geometrically speaking, these *torculi* are akin to toruses; that is, surfaces of revolution generated by the rotation of a circumference around an axis that lies in the plane of the circle but does not intersect it. During the sixteenth century, a number of torus vaults were built in ashlar masonry in Spain; the most significant ones are the vault in the inner chamber or *recapilla* of the chapel of Gil Rodríguez de Junterón in the Cathedral of Murcia (see fig. 2) and the annular vault around the courtyard of the palace of Charles the V in Granada.

Fig. 2. Chapel of Gil Rodríguez de Junterón in the Cathedral of Murcia. 1525-c. 1545. Photograph by David Frutos

A number of Italian or Italianate figurative artists were connected to these vaults, at least in the first design stages. Gil Rodríguez de Junterón asked for permission to build his chapel on March 27, 1525. At that moment, the master mason in the Cathedral of Murcia was Jacopo Torni, known in Italy as *L'Indaco vecchio* and in Spain as *Jacobo Florentino*, a painter trained in Ghirlandaio's workshop; later on, he worked with Pinturicchio and Michelangelo. Torni died in 1526, so he cannot have carried substantive work in the chapel; instead, the supervision of the works was probably carried out by Torni's successor, Jerónimo Quijano, since he was listed as a creditor for the chapel's altarpiece in Junteron's will [Gutiérrez-Cortines 1987: 164-167; Villela 1999]. As for the palace of Charles the V, the general plan and the direction of the first phases of construction were entrusted to Pedro Machuca, a painter who had stayed in Italy for a number of years, probably in Raphael's circle. However, Pedro Machuca died in 1550 and the annular vault was built by his son Luis between 1562 and 1569 [Rosenthal 1985: 96-97, 116-118]. Although it is usually taken for granted that Luis Machuca had made a trip to Italy, the circumstances of this journey are not clear.

It is also worthwhile to remark that both polyhedral rings and torus vaults seem to have played a certain role as icons of geometry, perspective and stereotomy. We have already mentioned the ring in the panelling of the cabinet of Federico de Montefeltro in Urbino, placed among the symbols of the liberal arts as an emblem of geometry. Daniel Barbaro included as chapter headings in *La prattica della prospettiva* a fair number of toruses of different sorts, both in perspective and orthographic projection [Barbaro 1569:

3, 25, 43, 129, 159, 197] (fig. 3). The title page of one of the extant copies of Alonso de Vandelvira's stonecutting manual, *Libro de trazas de cortes de piedras*, prepared by Felipe Lázaro de Goiti [Vandelvira 1646], includes two stereotomic tracings,[2] placed symmetrically at both sides of a central panel, and a third one below it (fig. 4). The tracing at the right of the central panel is a general scheme that can represent a number of stereotomic problems solved by squaring; by contrast, the motif at the left depicts the template-construction method for a horizontal-axis torus vault. Thus, the torus vault plays the role of an emblem of one of the basic methods in stonecutting, just as toruses act as icons of geometry and perspective in Urbino panelling or Barbaro's treatise. All this suggests that Quattrocento polyhedral rings and sixteenth-century Spanish torus vaults can furnish an interesting case study on the connections between perspective and stereotomy and, in a more general way, between science, architecture and the figurative arts in the Early Modern period, both in Italy and Spain.

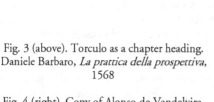

Fig. 3 (above). Torculo as a chapter heading. Daniele Barbaro, *La prattica della prospettiva*, 1568

Fig. 4 (right). Copy of Alonso de Vandelvira, *Libro de trazas de cortes de piedras*, by Felipe Lázaro de Goiti. 1646. Detail of frontispiece

Polyhedral rings as an exercise on perspective

According to a well known passage in Vasari [1568: 269], Paolo Uccello used to spend days and weeks alone, "*solo e quasi selvatico*", solving beautiful and difficult perspective problems about rotating arches and ribs, scaffoldings, or round columns placed at building corners, which brought him poverty rather than fame. One day he showed Donatello a number of sketches of rings depicted in perspective from different angles and a 72-face ball; the sculptor felt sorry for his friend and replied that these perspective exercises were useful only to marquetry specialists.

A number of drawings of toruses in the Uffizi, such as 1756A (fig. 5) and 1757A (fig. 6), have long been ascribed to Uccello [Kern 1915], taking into account these passages, and also the actual ring in Uccello's *Flood* in the Chiostro Verde in Santa Maria Novella (see fig. 10 below). However, this attribution has been contested recently [Roccasecca 1998; Roccasecca 2001c], at least for 1757A, ascribed by Roccasecca to the Sangallo circle, together with another torus drawing in the Uffizi, 832Ar. As for the *Chalice*, Roccasecca [2000] suggests a later date, in the seventeenth century at the circle of Evangelista Torricelli and the Accademia del Disegno in Florence. Thus, we will leave aside these drawings for the moment and come back to them after dealing with Piero's methods.

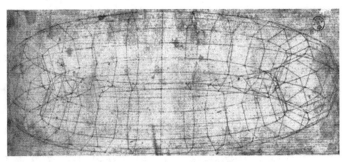
Fig. 4. Drawing of a mazzocchio. Uffizi 1756 A

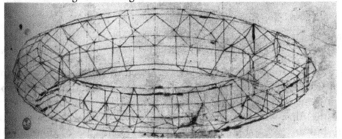
Fig. 5. Drawing of a mazzocchio. Uffizi 1757 A

At the beginning of the third book of Piero della Francesca's *De prospectiva pingendi*, after a number of introductory problems, Piero explains in great detail how to construct a perspective of a torus with eight horizontal parallels, divided in twelve sections by means of meridians [Della Francesca c. 1480: 37v-41v; Field 2005: 168].[3] As in drawings 1756 A and 1756 B, Piero substitutes an octagon for the meridian, while he depicts the parallels as circles, at least in the first stages.[4]

He starts his construction preparing a plan (fig. 7) and a profile (fig. 8) of the torus; he includes also in both drawings the station point and a line representing the picture plane of the perspective he is planning to draw [cf. Talbot 2006, suggesting the use of plan and elevation in the Uffizi *Chalice*]. To begin with, he constructs the cross-section of the torus in the shape of an octagon, inscribing it in a square "*de la quantità que tu vole fare grosso il torculo*", that is, the lesser diameter of the torus [Della Francesca c. 1480: 37v];[5] this octagon, drawn in the profile, will act as a generatrix or meridian of the torus. Next, he constructs a number of circular parallels or directrixes of the surface in the plan, taking into account "*la quantità che tu intendi fare grande il torculo*", that is, the larger radius of the torus, and the horizontal projections of the corners of the octagon, which he transfers from the profile to the plan using the compass. He explains dutifully that since the ring is "*giacente piano*", that is, lying horizontally, he only needs to trace four parallels. However, in the next step, he divides the parallels in twelve parts; each of the twelve division points gives the position of a meridian.

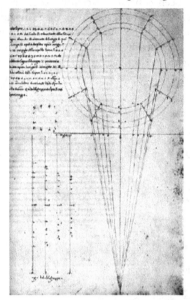

Fig. 7. Plan of a mazzocchio. Piero della Francesca, *De prospettiva pingendi*, c. 1475

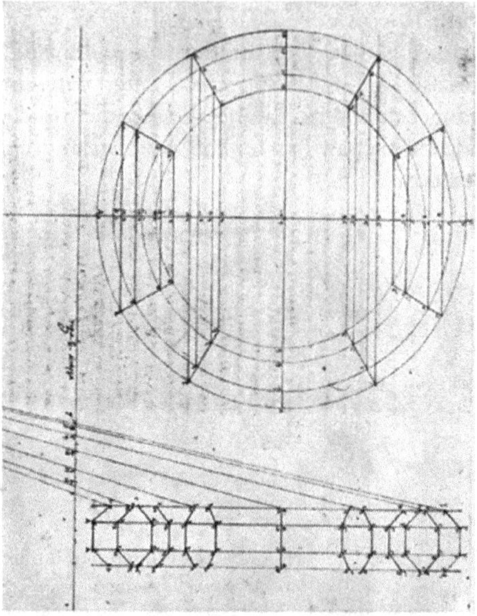

Fig. 8. Profile and plan of a mazzocchio. Piero della Francesca, *De prospettiva pingendi*, c. 1475

As a result of this construction, the plan is placed in top of the profile, in the same sheet of paper (fig. 8). The use of double orthographic projection and the superposition of elements in plan, elevation and profile are characteristic traits of masons' tracings, at least north of the Alps; when Dürer [1525: 84v][6] explains how to prepare a perspective using a plan and an elevation of a cube, he ascribes the use of double orthographic projection to masons. In fact, transfers between orthographic views play a central role in the method used by Mathes Roriczer to construct an elevation starting from a plan, the "secret" of late mediaeval masons [Frankl 1945; Shelby 1977]. In Roriczer's booklet about the geometric control of pinnacles, *Büchlein von der Fialen Gerechtigkeit*, [1486: 5r-8r] the issue involves two different problems: first, how to determine the height of a given member, following a number of geometrical rules that rely for the most part in rotated squares; second, how to maintain the correlation between the horizontal and vertical projections of each point. In manuals of descriptive geometry, as well as in stonecutter's drawings from the mid-sixteenth century on, this correlation is assured by the use of reference lines connecting the horizontal and vertical projections of each point. However, Roriczer does not use reference lines, but rather transfers measures from the plan to the elevation, using as a guide a vertical line placed at the axis of symmetry of his pinnacle, much in the same way as Piero does with his ring.

Roriczer's explanation is far from clear; possibly this is why Pérouse de Montclos [2001: 184-185] surmised that masons could not have taken the method from Roriczer, besides the obvious fact that the great majority of French and Spanish stonemasons at the beginning of the sixteenth century could not read German; in fact, a fair number could not read in their mother tongue [Marías 1989: 476].

Of course, Piero could not have taken his method from Roriczer, since the *Booklet on Pinnacles*, which in any case enjoyed little circulation, was published in 1486, when Piero was blind; nor could Roriczer have taken his method from Piero, because *De prospettiva pingendi* was not published before 1899. Rather, we should consider that both Piero and Roriczer were using a standard drafting method; references to such practices appear in *De re aedificatoria* and a well-known letter to Leo X, attributed to Raphael and Baldassare Castiglione [Alberti 1966: 99; Sanzio c. 1510; see also Lotz 1977;[7] Toker 1985; Di Teodoro 2001; Di Teodoro 2002; Di Teodoro 2003]. Besides, Piero [Della Francesca c. 1480: 38v] is in fact reversing the stonecutters' typical method and extracting the plan from the profile, although in a later step he will construct the meridians in the profile starting from their horizontal projections, thus extracting the side elevation from the plan.

While jumping from plan to profile and the other way around, Piero painstakingly places a string over a line joining each point in the torus to the station point, 96 points in total; at the same time, he transfers the intersection points of these lines with the picture plane to a set of waxed paper rulers or *rige*. Of course, each of these lines is a ray of the visual pyramid, to put it in Albertian terms,[8] and the intersection of each of these rays with the picture plane furnishes the perspective of a point of the torus. In order to draw the final perspective (fig. 9), the plan allows Piero to determine whether the perspective of a particular point in the ring falls to the right or to the left of a vertical line passing through the center point; however, it gives no information about the height of any point. Thus, in order to compute the height of a point in the perspective, Piero uses the profile, which allows measuring this height easily; in turn, the profile gives no information about the horizontal position of the point.

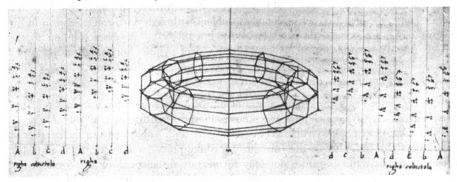

Fig. 9. Perspective of a mazzocchio. Piero della Francesca, *De prospettiva pingendi*, c. 1475

Piero transfers these points, both from the plan and the profile, to a separate sheet of paper by means of the waxed paper rulers and another set of wooden rulers. In fact, these rulers are used as a gauge, a typical stonecutters' instrument. Piero instructs the reader to make marks on the ruler measuring the distances for each point to a vertical line passing through the centric point, much in the same way as stonemasons used their gauges to transfer measures from their tracings to the stones that were to be dressed [Bechmann 1993: 61; Calvo 1999: I,132-133].

It may seem surprising to see Piero engaging in such complex operations, instead of using simpler techniques, such as vanishing and distance points. Piero himself gives two different reasons in the introduction to the third book of *De prospectiva pingendi* [Della Francesca c. 1480: 32r]. He deals in this book with solids bounded by different surfaces and "*diversamenti posti*", that is, rotated with regard to the picture plane. To tackle these difficult problems, he introduces in this book a new method, which will be easier to explain and understand than those used in the first two books in his treatise, while allowing him to avoid "the great multitude of lines that these bodies would need if using the first method".

This last explanation does not sound convincing; in order to draw the ring the draftsman should trace 96 lines[9] and transfer the intersections of these lines to the perspective. However, we must take into account that a number of perspective techniques, now familiar to us, were not known at Piero's time. The "distance vanishing point" [see García-Salgado 2003],[10] that is, one of a pair of points placed at the horizon line so that their distances to the centric point are equal to the distance between station point and perspective plane, is not used as an operative method in Piero's treatise, as Field [2005: 148-150] has remarked. Although his use of the diagonal of a square to compute the position of transversal lines in books I and II of *De prospectiva pingendi* is a significant leap in this direction, he still needed to construct an enclosing rectangle using projection and intersection to trace the diagonal, which in fact does not reach the horizon line and the distance vanishing point [Della Francesca c. 1480: 7r-17 r; see also Panofsky 1927; Klein 1961; Vagnetti 1980; Field 2005, 148-150].[11]

Besides, Piero did not know a general method for the placement of the vanishing point of oblique straight lines lying in a horizontal plane at a given angle to the picture plane. Of course, he knew the solution for a particular case of this problem; namely, the vanishing point of orthogonals, that is, lines forming an angle of 90° to the picture plane, which is the Albertian centric point. In a particular occasion he used the "distance vanishing point", that is, the vanishing point of lines forming an angle of 45° to the picture plane, to solve a rather far-fetched problem, although this is an exception in his treatise and not a general method [Field 2005: 148-150]. The exact solution to the general case of the problem of the vanishing point of oblique horizontal lines was put forward only by Guidobaldo del Monte [1600: 43-44; see also Kemp: 1990, 89-90]; Piero mentions nothing of the sort.

On the other hand, the empirical use of the "distance vanishing point" to put into perspective a set of parallel lines was known in the Quattrocento, as the *sinopia* of Uccello's *Nativity* in San Martino alla Scala suggests, and is ubiquitous in the treatises of Pelerin and Cousin. That is, the "distance vanishing point" was frequently used in the fifteenth and sixteenth centuries not only to measure distances along orthogonals, but also as a "vanishing point" when drawing horizontal lines lying at an angle of 45° to the picture plane, just as the centric point is used as the "vanishing point" of the orthogonals to the picture plane [Klein 1961; Sindona 1980; Kemp: 1990, 37-38; Raggio 1996; Roccasecca 2001a; Pelerin 1505: VIIr, VIIIr; VIIIv, XVIIr, XXr, XXv, XXIr, XXVIr; Serlio 1545: 60r-62v; Cousin 1560: 7r, 8r, 8v, 26 r, etc].

However, these empirical solutions are not mentioned in *De prospectiva pingendi*. In fact, the layout of Piero's ring, with twelve meridians, seems to be chosen to put forward a case that cannot be solved by these methods. A ring with eight parallels and eight meridians, an obvious choice, could have been solved with two sets of orthogonals, two sets of transversals and four sets of lines meeting at both "distance vanishing points", a notion used, at least empirically, by Ucello in San Martino alla Scala. By contrast, a twelve-sided ring cannot be solved through these empirical methods, showing thus the power of Piero's "other method", which is even more striking taking into account that Del Monte's construction was not known in the period.

All this means that Piero knew a general perspective method, based on orthogonal projections and explained in his third book, that was powerful enough to tackle any perspectival problem in Piero's view. He also dealt in the first and second books of *De prospectiva pingendi* with a number of specific techniques, such as the use of orthogonals meeting at the centric point, the diagonal of a square to transfer measures from a transverse line to an orthogonal, and auxiliary vertical planes to compute the height of a point above the ground plane. These methods were useful for constructing simple figures, such as polygons, cubes, columns, wells, pedestals and even groin vaults. However, when confronted to complex figures, starting from the torus or polyhedral ring and progressing to rotated cubes, vaults, capitals and human faces, the general method showed all his power.

According to Roccasecca [1998], the ring at the Louvre presents pricked holes on most corners of the figure, as well as a fair number of lines incised with a stylus, all of them underlying the actual drawn lines. This suggests that the drawing could have been prepared using Piero's technique, since the rulers should make any auxiliary construction unnecessary. However, the perspectival methods applied to this problem seem to have evolved during the first half of the sixteenth century. In the Uffizi drawings of rings, as well as in the *Chalice* [Kern 1915; Roccasecca 1998; Talbot 2006], a number of incised lines are placed at the intersection of the meridian planes with the horizontal planes for each parallel. Of course, for each parallel, all these lines intersect at the center of the parallel; these centers are placed at the axis of the torus, at different heights. All this suggests that the draftsmen of these rings, quite probably in the Sangallo circle for 1737A and 832 Ar, and possibly for 1736A, used a different technique, based in the rotation of a horizontal plane to place it at a frontal position, thus allowing the parallel to be constructed in true shape and then transferred to the perspective.[12]

Thus, in the fifteenth century Piero solved the problem of the *torculo* using projections from plan and elevation, while in the sixteenth century some draftsmen, probably from the Sangallo circle, complemented the use of plan and elevation with sophisticated methods to tackle this difficult problem again. All this explains the role played by the polyhedral ring as an emblem of Quattrocento perspective; it allowed the real masters of perspective to show all their cunning.

Polyhedral rings in Italian Renaissance painting and marquetry

The rings drawn in all the studies we have seen so far share a common trait: the axis of revolution of the surface is vertical; in Piero's terms [Della Francesca c. 1480: 37v], the figure is set lying on a plane or "*giacente piano*". As a result, during surface generation, each point describes a circle lying in a horizontal plane; thus, at the first stages of the perspective construction, while tracing the plan of the ring, the draftsman can use the compass to draw easily the plan of each of these circles, as Piero states clearly [Della Francesca c. 1480: 37v].[13] Also, in this position the torus has a horizontal plane of symmetry, and thus each of the eight horizontal sections of the ring lies above another section; as a consequence, the draftsman only needs to draw four circles and use four rulers [Della Francesca c. 1480: 37v].[14]

However, most extant polyhedral rings in Quattrocento paintings and marquetry panels, such as all three panels of Paolo Uccello's *Battle of San Romano*, the *Flood* in the Chiostro Verde in Santa Maria Novella, also by Uccello (fig. 10), or the inlaid panels in the Urbino cabinet, do not lie in horizontal planes.[15] Thus, each point in the parallels will describe a circle in a sloping plane during surface generation, and the parallels will be projected as ellipses in the plan.

Fig. 7. Paolo Uccello. Detail of the torus from the *Flood.* Chiostro Verde, Santa Maria Novella, Florence

This simple detail poses a formidable problem for the draftsman, right at the beginning of the perspective construction. To begin with, the notion of the ellipse was not widespread in the fifteenth century; the first translation of Apollonius of Perga's *Conics*, by Federigo Commandino, was not published until 1566. It is true that the construction of conic sections explained by Dürer [1525: 16r; see also Roccasecca 1998] was known in the Sangallo circles and that Spanish stonemasons found an empirical way to construct the projections of circles lying in sloping planes along the sixteenth century, as we shall see later, but Quattrocento literature mentions nothing of the sort.[16]

Piero was aware of this problem; in fact, he states explicitly that "We shall put the *torculo* lying flat; however, when lying in another way, it will be necessary to make as much rulers as circles are contained in the *torculo*".[17] That is, Piero is considering the case of the sloping ring, advising the reader that in this case the circles do not stand one above the other, since there is no horizontal symmetry plane, and instructing the reader to use as many rulers as circles are contained in the figure.

Piero does not explain in so many words how to construct this sloping ring, but the next section in *De prospectiva pingendi* deals with a rotated cube [Della Francesca c. 1480: 42r-43v; see also Evans 1995: 153-154]. The position of this section within the manuscript is somewhat surprising, since it stands between the flat-lying ring and the base of the column, which includes a fair number of flat-lying toruses.[18] Thus, the placement of the rotated cube in the text of *De prospectiva pingendi* hints strongly that

Piero included it as a simplified demonstration of the method he was proposing to solve the problem of the rotated torus.

In order to rotate the cube so that "*nisuno suo lato sia equidistante al termine posto*", that is, that no face or edge or the cube would lie parallel to the picture plane, Piero follows a complex, although quite logical, procedure in four steps [Della Francesca c. 1480: 42r-43v]. First, he draws in plan and elevation a cube lying on a horizontal plane (fig. 11); however, the plan of the cube is rotated with regard to the reference lines that connect plan and elevation; thus, the elevation shows an oblique view of the cube. In the second step (also shown in fig. 11), Piero rotates the elevation of the cube around a horizontal axis passing through one of its lower corners. Piero easily manages to construct the plan of the rotated cube, taking into account that each point will move during the rotation on a plane that is orthogonal to the rotation axis, that is, a frontal plane; also, each point of the plan will be joined to the rotated elevation by a reference line.[19] That is, once more Piero reverses the typical method of mediaeval masons and extracts the plan from the elevation.

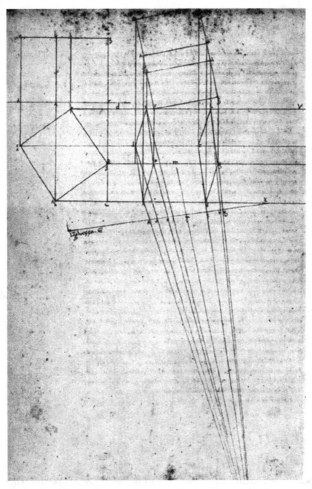

Fig. 11. Plan and elevation of a flat-lying and a rotated cube. Piero della Francesca, *De prospettiva pingendi*, c. 1475

However, the cube is not yet in its final position, since Piero intends that "no edge should be parallel to the picture plane", as we have seen; in order to do that, he simply rotates the picture plane of the perspective. Now, he can place the point of view of the perspective and trace all the visual rays in order to compute the horizontal position of each point in the perspective, as he did with the polyhedral ring (still in fig. 11).

Once this is done, he cannot perform directly the same operation starting from a vertical projection, since he has still no profile, but only an elevation, which is not even parallel to the picture plane of the perspective. Thus, Piero constructs on a separate sheet (fig. 12) a profile, that is, an orthogonal projection on a plane that is perpendicular to the picture plane of the perspective. In order to do so, he starts from the rotated plan and transfers the height of each point from the rotated elevation to the profile; once this is done, he can trace visual rays in the profile view, find their intersections with the picture plane and transfer the resulting vertical positions to the perspective in order to complete it (fig. 13).

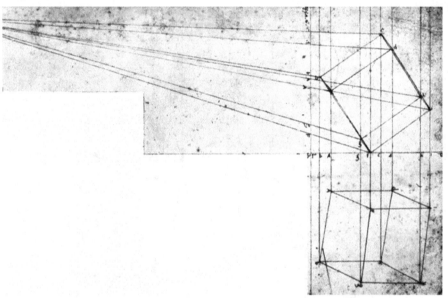

Fig. 12. Plan and profile of a rotated cube. Piero della Francesca, Fig. 12. Plan and profile of a rotated cube. Piero della Francesca, *De prospettiva pingendi*, c. 1475

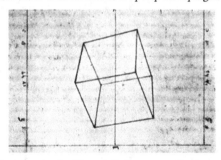

Fig. 13. Perspective of a rotated cube. Piero della Francesca, *De prospettiva pingendi*, c. 1475

This method can be easily extrapolated to the polyhedral ring, although the number of lines and point transfers involved is quite daunting (fig. 14). It can be also used to rotate heads, as Robin Evans [1995: 153-154] has suggested; although Piero does not explain the technique, Albrecht Dürer does so in *De symmetria partium in rectis formis humanorum corporum* [Dürer 1532: 117v-118v, 126v-127r, 130v, 132r-134r].[20]

All this brings us back to Vasari's [1568:269] mysterious comment on Uccello "rotating arches and ribs". Both Piero [Della Francesca c. 1480: 29r-30r; see also Kemp 1990: 29] and Sebastiano Serlio [1545: 45v-46r] explain how to construct the perspective of a groin vault (fig. 15), but they make no use of rotations of any kind.[21] By contrast, these rib rotations were standard practice in late Gothic stonecutting practice. Hernán Ruiz [c. 1550: 46v], Philibert de L'Orme [1567: 108v], Alonso de Vandelvira [c. 1580: 96v], Alonso de Guardia [c. 1600: 85b], Gelabert [1977: 281] and a good many other Renaissance writers on stonecutting explain how to compute the curvature of the ribs of a tierceron vault and the height of the secondary keystones (fig. 16), rotating diagonal ribs and tiercerons around a vertical axis, in order to bring them to the plane of the transverse arches and depict them as circular arcs (see also [Rabasa 1996]).

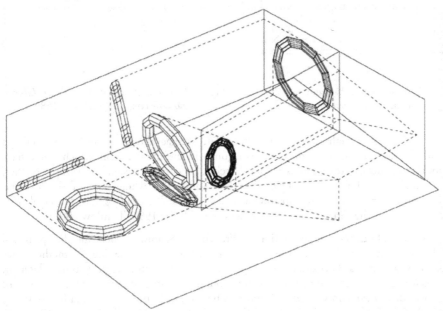

Fig. 14. Piero's construction for a rotated cube applied to a mazzocchio. For the sake of clarity, only the extreme instances of each set of reference and projecting lines are shown.
Drawing by José Calvo-López

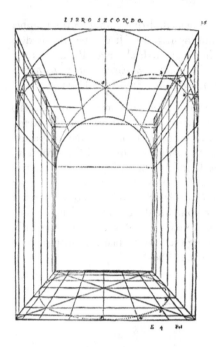
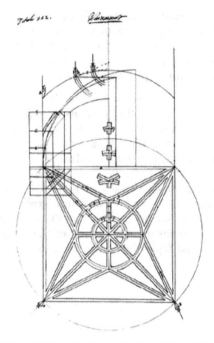

Fig. 15. Groin vault. Sebastiano Serlio. *Tutte l'opere di architettura*, 1600

Fig. 16. Rib vault. Alonso de Vandelvira, *Libro de trazas de cortes de piedras*, c. 1580

This ingenious method, quite probably of late Mediaeval origin, avoids constructing a standard orthographic projection of the diagonal rib and the tierceron, which would have appeared in such a projection as elliptical arches.[22] German masters went much further, since they rotated not only single ribs; they unfolded a whole path of ribs in their *Netzgewolbe* or net vaults, to construct a virtual rib, the *Prinzipalbogen* or principal arch, extending from the springer to the main keystone [Müller 1990; Tomlow 2009]

However, both the simple rotations of French and Spanish masters and the "principal arch" technique involve rotations around a vertical axis, while Piero's method uses rotations around a horizontal axis. Rotations of this kind appear much later in stonecutting. An interesting example can be found in another stonecutting manuscript of the period, *Cerramientos y trazas de montea*, by Ginés Martínez de Aranda [c. 1600: 11, 15, 40, 46]; the author rotates the intrados joint of a number of skew arches and rear-arches in order to compute the angle between the intrados joint and the face joint. At that moment, however, both horizontal and vertical axis rotations were standard practice in perspective treatises, as Jean Cousin [1560: 27r, 28r, 32 r, etc.] makes clear.

All this refers to the rotation of the objects depicted in orthographic projections, while projection planes stand still. By contrast, when Piero constructs the rotated profile of his cube, the last step in the perspective construction process, he is performing a typical operation of descriptive geometry, known much later as change of projection plane [Della Francesca, c. 1480: 43r-43v; see also La Gournerie 1860: VI-VII]. That is, he leaves the cube standing still and rotates the projection plane of the elevation in order to construct a profile. This operation is quite usual in French and Spanish stonecutting treatises and manuscripts of the sixteenth century; it appears in Hernán Ruiz [c. 1550:

47v], Philibert de L'Orme [1567: 113r], Alonso de Vandelvira [c. 1580: 24v, 80r, 81r, 90r, 90v, etc.], Ginés Martínez de Aranda [c. 1600: 7, 9, 10, 72, 76, 78, 86, etc.], and others, and even in De L'Orme's carpentry treatise, the *Nouvelles inventions pour bien bastir et a petits frais* [1561: 13r]. Excepting this later treatise,[23] all stonecutting texts use the same method employed by Piero; the mason should trace reference lines from all relevant points in the plan, and once this is done he should transfer the heights of each point in the original elevation to the new elevation or, in Piero's case, to the new profile.

Thus, it seems clear that Piero and other authors on perspective, such as Cousin, shared with stonecutting literature not only the use of orthographic projections, but also a number of rather sophisticated techniques that much later were to find a place in descriptive geometry, such as rotations and change of projection planes. We shall tackle the problems of the formation of these techniques and the direction of influences between perspective and stonecutting in the final section of this article, once we have dealt with the stone rings of the Spanish Renaissance.

Torus vaults in Spanish Renaissance architectural practice

The oldest of these built toruses is the vault of the inner chapel or *recapilla* of Gil Rodríguez de Junterón in the Cathedral of Murcia (figs. 2, 17, 18), begun in 1525 and completed around 1543 [González Simancas 1905: 151-152; Gutiérrez-Cortines 1987: 164, 170]. Junterón had spent some years in Rome during the papacy of Julius II, serving in the curia as protonotary. When he asked for permission to build his chapel on March 27, 1525, the master mason in the cathedral was Jacopo Torni, a Florentine painter apprenticed in Ghirlandaio's workshop.

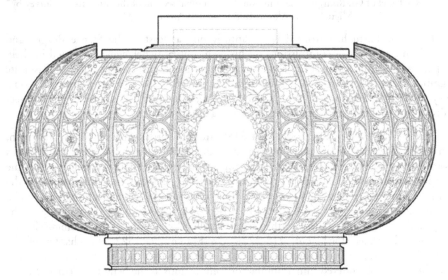

Fig. 17. Chapel of Gil Rodríguez de Junterón. Plan of the inner room. Survey by Miguel Ángel Alonso and Ana López Mozo

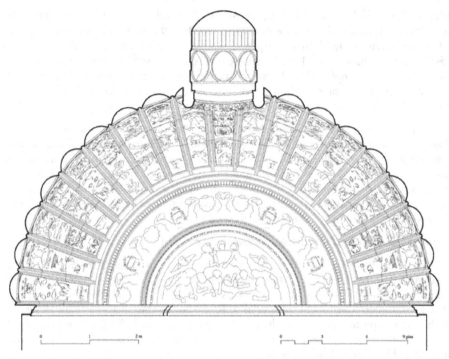

Fig. 18. Chapel of Gil Rodríguez de Junterón. Longitudinal section of the inner room. Survey by Miguel Ángel Alonso-Rodríguez and Ana López-Mozo

Later on, Torni had worked with Pinturicchio, at least at the Stanze Borgia, and carried on a number of commissions on his own, until Michelangelo requested his help to paint the Sistine ceiling. His brother may have worked with the Sangallos; this is quite significant for our purposes, taking into account Roccasecca's attribution of some polyhedral ring drawings to the Sangallos' circle, although the connection is indirect [Vera 1993:16-17; Roccasecca 1998; Roccasecca 2002b].

Vasari [1550: 528-529; 1568: 524-525] stresses Torni's qualities as a draftsman, while pointing out his laziness; however, his short Spanish career suggests that Vasari's portrait may be biased. He arrived in Spain not later than 1520, starting work at the Royal Chapel in the Cathedral of Granada, were he received the commision for the Annunciation group over the vestry door, three panels for the Santa Cruz altarpiece and a fair number of decorative works; he also carried on a restoration of the painting of the Virgen de la Antigua in the Cathedral of Seville [Velasco c. 1564; Gómez-Moreno 1925b; Calvo Castellón 1994, 218-222]. On March 29, 1522, the chapter of the Cathedral of Murcia asked him to accept the post of master mason of the cathedral; at the same time they offered him a number of commissions at the behest of Pedro Fajardo, first Marquis of Vélez, such as the main altarpiece in the cathedral and a number of sculptures for his family chapel [Gutiérrez-Cortines 1987: 64-65; González Simancas 1905: 87]. His main duty as master mason was the supervision of the great bell-tower of the cathedral, whose construction had begun three years before under the direction of another Italian artist, known as Francisco Florentino. In less than four years, Torni completed the first story of the tower, which includes an ornate sacristy with two remarkable stereotomical pieces, a sail vault and a skew passage, while holding the post of master mason of San Jerónimo in Granada before his death in Villena, probably when inspecting the construction of some works in the church of Santiago or its parish house [Gutiérrez-Cortines 1987: 61-66, 112-129, 135-136; Calvo 2005a].

Although no documentary evidence connects Torni with Junterón's chapel, historians have traditionally attributed the general design of the chapel to him, taking into account a number of Italianate traits in the chapel and some similarities with the cathedral sacristy and other documented work by Torni. We should take also into account that Junterón seems to have been a political client of Fajardo, since he undertook a difficult mission in the war of the Comuneros on his behalf [González Simancas, 1905: 156-157; Gutiérrez-Cortines 1987: 161-164].

In any case, since Torni died ten months after the beginning of the chapel's construction, it is clear that he was not in charge of the actual execution of the chapel. This task was quite probably undertaken by Jerónimo Quijano, a Spanish sculptor trained in the entourage of Philippe Vigarny. He had carried on works with Torni and his father-in-law, Juan López de Velasco, and held the post of master mason in the cathedral after Torni's death; much later, he was listed as a creditor in Junterón's will for the altarpiece in the chapel [González Simancas 1905: 156-157; Gutiérrez-Cortines 1987: 164; Villella 2002].

Cristina Gutiérrez-Cortines and Marzia Villella have stressed the traits that connect Junterón's inner chapel with the first project of Julius II's tomb, such as the general plan, a rectangle terminating in two semicircles or the use of niches between pilasters [Gutiérrez-Cortines 1987: 176-177; Villella 1999]. However, the most singular member in the chapel, the vault over the inner room, has no connection with Julius II's tomb nor with Quijano's other work. When confronted with a similar problem in Santa María in Chinchilla, Quijano used a straightforward solution: a barrel vault terminating in two quarter-sphere vaults [Gutiérrez-Cortines 1987: 200-211; Calvo 2005b; Calvo 2008].

By contrast, the vault over Junterón's inner vault is a horizontal axis torus vault; so to speak, one of the rings in the heads of the contenders in the Battle of San Romano or around the neck of the survivors of the Flood has completed its rotation, reaching a vertical plane. However, only the fourth part of the surface is actually built, since the upper half of the outer part of the surface is enough to span the chapel area. To be precise, less than half a quarter of the entire surface is used; in fact, the survey by Miguel Ángel Alonso and Ana López Mozo (figs. 17, 18) has shown that the generatrix is somewhat smaller than a semicircle, in order to increase the width of the arch around the relief of the *Paradise* in the tympanum of the back wall of the chapel and the span of the entrance arch. The treatment of this surface shows a fair understanding of the geometry of this figure. Instead of giving the same value to meridians and parallels, as Uccello or the cabinet-makers at Urbino had done, the builder of the chapel stressed the radial pattern of the generatrices, thus bringing the focus of the whole architectural composition of the chapel to the reliefs of the *Paradise* and the *Nativity* in the chapel's altarpiece. In contrast, the builder broke the continuity of the directrixes at both sides of each generatrix; in this way, the directrixes adopt an asymmetric pattern, made almost invisible by the rich array of grotesque sculpture in each compartment (fig. 19). At the same time, the portion of the vault included between two consecutive generatrices acts as a course, and the breaking of the directrixes prevents the voussoir from slipping, a sound constructive practice [Calvo 2005c: 152, 161-163].

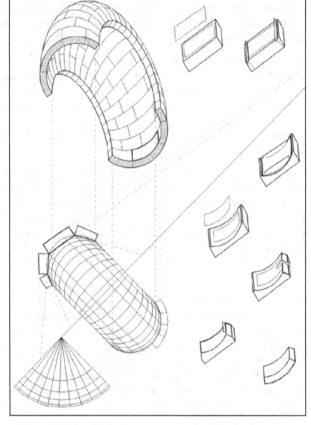

Fig. 19. Vault over the inner room in the Chapel of Junterón.
Stonecutting diagram by José Calvo-López

We shall come back to these issues in the next section; for the moment, we shall discuss another solution for the torus vault built over the lower gallery around the courtyard of the palace of Charles V in Granada. Quite remarkably, this piece shows a number or parallels with the inner chapel of Junterón, while choosing opposite solutions in a number of crucial points. A wide consensus supports the authorship of Pedro Machuca for the project of the building, although another architect, Luis de Vega, was consulted and perhaps introduced a number of changes; in contrast, Manfredo Tafuri has argued pervasively for a project by the hand of Giulio Romano, although this hypothesis is grounded on stylistic terms only [Rosenthal 1985: 11-12; Tafuri 1987].

Lázaro de Velasco, the Spanish son of Jacopo Torni, mentions an Italian sojourn of Pedro Machuca in the introduction to his translation of Vitruvius, remarking that he arrived more or less at the same time as Torni [Velasco c. 1564]. However, nothing is known about Machuca's Italian activity. Taking his cue from Velasco, Earl Rosenthal [1985:16-17] has speculated with the possibility that Machuca may have frequented in Italy the same circles as Jacopo Torni. However, this hypothesis is at odds with the general Raphaelesque character of Machuca's paintings and the stylistic details that connect his architectural work with Giulio Romano. On the other hand, Machuca was in fact connected with Torni shortly after they arrived in Spain; each of them painted three panels for the Santa Cruz altarpiece, flanking the central piece by Dierick Bouts [Gómez Moreno 1925b; Calvo Castellón 1994: 218-222].

Further, both Machuca and Torni were connected to the Mendoza family, whose patronage was crucial in the introduction of the Renaissance in Spain. Machuca directed the construction of the palace in the Alhambra as *escudero* or client of Luis Hurtado de Mendoza, a prominent member of the family, while Torni worked in the chapel of Nuestra Señora de la Antigua in the Cathedral of Seville, the funerary chapel of Diego Hurtado de Mendoza [Rosenthal 1985: 12-13; Velasco c. 1564; Morales 1992: 185-187]. Both Diego Hurtado de Mendoza and Pedro Fajardo, the political patron of Junterón and artistic patron of Torni, had been educated by Pietro Martire d'Angheria, an Italian humanist under the patronage of the father of the Captain-General of the Alhambra, the first Marquis of Mondéjar, Íñigo López de Mendoza, father of Luis Hurtado. It is also worthwhile to remark that Angheria has been included in the group of humanists that suggested to Charles the V that he build the palace, along with Baldassare Castiglione.[24] Besides, two prominent members of this group of humanists in the entourage of Charles V during his stay in Granada, Alfonso and Juan de Valdés, were to belong to the chapter of the Cathedral of Murcia later on [López de Toro 1953: 64-66, 77-78, 82-84, 198, 200, 202-205, 211-212, 276-277, 386-387, 413-415, 421-422; López de Toro 1955: 30-32, 258-259; 268-269; Rosenthal 1985: 10; Rosenthal 1988; Meseguer 1957]. It is also noteworthy that an important foreman of the building team in the palace of Granada, Juan de Marquina, had worked with Torni's predecessor in the Cathedral of Murcia, Francisco Florentino, and had lived in Murcia during the 1520's; he was probably aware of Junteron's plans for his inner chapel [Gutiérrez-Cortines 1987: 52, 61, 328-330; see also p. 95, note 111 and p. 356, note 141; Rosenthal 1985: 53-54]. Thus, a densely woven net of connections between patrons, artists and builders ties together the chapel of Junterón and the palace in Granada.

The vault in the lower gallery also shares a parallel history with the chapel in Murcia. Laying aside Vega's proposal for half-columns and pillars, Pedro Machuca had laid out the foundations for a ring of free-standing columns and prepared a model of the palace, which included part of the colonnade. However, as in the case of Torni and the Murcia vault, Machuca did not live to see the vault erected; as we have seen, it was built by his son Luis between 1562 and 1569; at that moment, however, Marquina was either dead or retired [Rosenthal 1985: 99-100, 112-119; Rosenthal 1988].

In other aspects, the vault in Granada reverses a good number of design traits in the Murcia archetype. First of all, its axis of revolution is vertical, just like Piero's toruses or the rings around the Uffizi's *Chalice*. Thus, the generatrices or meridians are semicircles laid out on vertical planes, while the directrixes or parallels are horizontal circumferences. Following the same constructive logic as the Murcia vault, here the roles of these curves are reversed. The horizontal directrixes act as bed joints, and thus are almost uninterrupted along the huge circumference of the gallery, while the vertical meridians are broken to prevent the voussoirs from slipping. In another striking parallel with the Murcia vault, the generatrix is not strictly a circumference, since the first course of the vault in the outer side, springing from the corridor wall, is set at a lower height than the first course in the inner side, resting on the columns that separate the corridor from the courtyard. By contrast, the treatment of the surface in the Granada wall is completely opposite to the Murcia one; instead of being hidden below a host of grotesque sculpture, the vault in the palace displays a completely nude intrados, making a great display of the network of joints and predating the characteristic *nudité d'intrados* in seventeenth century French vaults [Pérouse de Montclos 2001:108-111; Etlin 2009].

Rosenthal cites a number of classical precedents for this vault, such as the annular vault at the seventh level of the Praeneste sanctuary, the Maritime theater in Hadrian's Villa and the deambulatory in Santa Constanza [1985: 217-218; see also Rosenthal 1988]. He also suggests that Pedro Machuca could have been thinking of another constructive solution for the vault, such as a concrete vault in the Roman tradition; that he finally adopted an ashlar vault under pressure from Luis de Vega and other Castilian advisors; that it was not easy for him to maintain his original idea of free-standing columns, since the Castilian consultants favoured pilasters and half-columns; and that Luis Machuca was planning to have the rough tufa surface of the vault plastered and covered by frescoes [1985: 118-119]. Except the discussion about free-standing columns or pilasters, which dates from around 1530 and is well documented, the rest are hypotheses, since the documents from the 1560s are scarce. However, it seems clear that there was a great concern about the constructive feasibility of the solution. Besides, in the eyes of everybody, except maybe the Machucas, the aristocratic ideal of magnificence was tied to the use of ashlar, and the courtiers would hardly accept another material for such a visible element in an Imperial palace [see for example Díez del Corral 1992 or Barbé 1990].

All this suggests that either Pedro or Luis Machuca could have conceived an annular vault starting from sources in Antiquity, direct or indirect; we should remember that Serlio's third book presents a quite expressive drawing of the interior of Santa Constanza, which was included in the Spanish translation by Francisco de Villalpando [Serlio 1544, XX; Serlio 1552, III-XIIv]. However, taking into account the number of connections between Granada and Murcia, it is rather puzzling that Rosenthal does not connect the Granada vault with the Junterón one, although he makes much of Machuca's relationship with Torni [Rosenthal 1985:13, 16-17]. At least, Junteron's vault must have played in Granada the role of a feasibility check; that is, the vault in Murcia made clear that it was possible to build such unusual shapes in ashlar. However, influences in the opposite direction cannot be discarded. Either the Valdés brothers, who belonged with Junterón to the Murcia chapter, or Pedro Fajardo, who kept up a lively correspondence in Latin with Pietro Martire d'Angheria, could have informed Junterón about the early projects of the Granada palace.

It is quite easy to understand why the vault in the Granada palace, about twenty meters in internal diameter, has no derivatives. The vault in Murcia has given birth to a modest number of heirs. It is almost literally split in half at the crossing of the nearby church of Santiago in Orihuela: two niches flanking the central space, probably built by Quijano, are covered by surfaces generated by a quarter of a circle rotating around a horizontal axis and thus, the result is the eighth part of a full torus [Gutiérrez-Cortines 1987: 260; Palacios 2003: 225; Calvo 2005c: 43]. The treatment of the surface lies halfway between the vault in Murcia and the one in Granada, since it is divided into coffers; thus, it eschews the rich grotesque sculpture of the Junterón chapel, but does not show the nudity of the intrados of the Alhambra vault. This eighth-of-a-torus surface is also used in a wall arch in the crossing of the church in the convent of San Francisco in Baeza, by Andrés de Vandelvira. This is quite interesting, since Alonso de Vandelvira, son of Andrés, wrote the only manuscript that deals with the construction of torus vaults; we shall deal with his explanations of steretomical procedures applied to torus vaults in the next section.

Usually, stonecutting treatises or manuscripts describe the problems they are going to deal with in geometrical or constructive terms. To point out an extreme example, Ginés Martínez de Aranda [c. 1600, 191-193] includes in *Cerramientos y trazas de montea* such an elaborate piece as the "Puerta en torre redonda contra capialzado cuadrado desquijado de arco en torre cavada", that is, a doorway in a convex wall confronting a square rear-arch joined to an arch in a concave wall. However, a small number of stonecutting problems, such as the *Vis de Saint-Gilles*, a spiral staircase covered by a sloping barrel vault or the *Arrière-voussure de Saint-Antoine*, a rear-arch with curved joints, take their names from the cities or places where a particularly significant example of each problem is built. These geographical denominations or *appellations d'origine*, in Pérouse de Montclos' words [2001, 202-204], are also used in Spain. Vandelvira's manuscript includes the *Ochavo de La Guardia*, a coffered octagonal vault; the *Caracol de Mallorca*, a spiral staircase with a winding newel; the *Bóveda de Cuenca*, a square coffered vault, and the *Bóveda de Murcia*, a horizontal axis torus vault, quite similar to the vault in the *recapilla* of Junterón [Vandelvira c. 1580: 103v, 51r, 97v, 69v].

We should stress two points about these appellations. First, the small number of geographical denominations indicates that these problems and built examples are seen as exceptional challenges, as we remarked when dealing with Goiti's treatment of the torus vault as an emblem of stonecutting. Second, the appellations are applied to geometrical problems, rather than to actual built examples. This is quite clear in France, where *Arrière-voussures de Marseille* or *Pendentifs de Valence* are not difficult to find. A small detail in Vandelvira's manuscript hints in this direction; when talking about the *Vis de Saint-Gilles* he points out that "it is executed in Saint-Gilles in France" and not "it is in Saint-Gilles".[25] This is essential for our purposes, since Vandelvira's drawing and explanation of the *Bóveda de Murcia* depart in a number of small but significant details from the built example in the chapel of Junterón [Vandelvira c. 1580, 52v, 69v].

Vandelvira starts his construction (fig. 20) by tracing the outline of the plan of the vault. He then divides the end semicircles in an odd number of voussoirs and draws a number of parallels of the surface, starting from the division points. This involves a first point of departure from the Murcia example, since the end sections of the springing of the vault in Junterón chapel are not exactly semicircles, as we have seen. Thus, it is clear that Vandelvira is not including in his manuscript either a survey or a project of the Murcia vault, but rather a general solution to the problem of horizontal-axis torus vaults; therefore, he leaves out a trait that stems from the adaptation of an abstract model to the particular circumstances of Junterón's chapel.[26]

Since the parallels in Vandelvira's tracing run from one end of the vault to the other, the joints are not broken between one course and the next, as in the actual Murcia vault. Of course, this layout does not prevent the voussoirs from slipping; Vandelvira is aware of the fact, since he makes a passing remark about an alternative solution involving "ligazones", that is, the breaking of joints between one course and the next, but he does not explain this solution in detail.

Once this is done, Vandelvira draws the elevation of the vault, tracing reference lines from the plan until they meet the springing line in the elevation. Since the intersections with this line allow Vandelvira to measure the diameters of the longitudinal section of the vault and each parallel, he can easily construct the section and the parallels. He can also divide the section in an odd number of voussoirs and trace a number of lines from the midpoint of the springing line to the division points; this operation furnishes an array of lines converging in the central points of the springing line, which represent the meridians of the vault.[27]

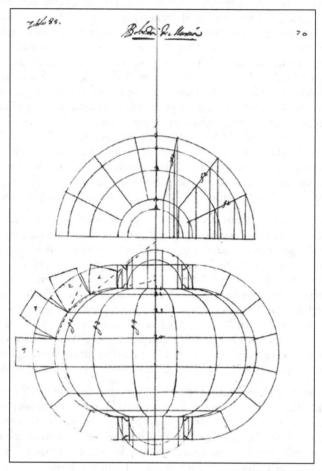

Fig. 20. Horizontal-axis torus vault. Alonso de Vandelvira,
Libro de trazas de cortes de piedras, c. 1580

Next, Vandelvira constructs the horizontal projections of these meridians in the plan. Although the meridians are semicircles, they are projected as half-ellipses, since they lie on sloping planes. To perform this complex operation, Vandelvira draws reference lines starting from the intersections of each meridian with each parallel in the elevation, and brings them to the plan; where each of these lines meets the corresponding parallel, he can place a point belonging to a meridian. Once all points of a meridian are placed, he joins them by means of arcs of a circle, taking them in groups of three points.[28] That is, here Vandelvira inverts the traditional Gothic procedure of extracting the elevation from the plan and places the horizontal projections of the points of the meridians starting from the elevation, much in the same way as Piero had done with the parallels in his ring or with his rotated cube.

The crucial steps in the process start at this point. Vandelvira has used orthogonal projections as the foundation of the stereotomic method; now he will apply a different geometrical technique, involving an approximate development of the torus surface, for the preparation of the intrados templates of the voussoirs. In order to construct these templates, Vandelvira uses a number of imaginary cones, acting as substitutes for the torus surface, which is non-developable. These cones have their vertexes placed at the axis of the torus; thus, each cone passes through two consecutive parallels. He first traces a generatrix of the cone, joining two consecutive division points in the springing line, carrying it over until the line meets the axis of the torus; the intersection point will be the vertex of the cone. Then, Vandelvira develops the cone using a well-known technique: he traces two circular arcs with their centres placed at the vertex of the cone; this will give him the edges of the intrados template of the voussoir corresponding to the parallels, while the generatrix stands for one of the meridians. In order to complete the development of the intrados face of the stone, Vandelvira should compute the angle between both developed meridians, taking into account the length of the circular arcs that represent the parallels; however, he makes no mention of this calculation, although he draws the final meridian.[29]

These cones seem to be indirectly related to Alberti's pyramid. When explaining the same procedure, Alonso de Guardia [c. 1600: 87v] advises the reader to "place the straightedge passing through two points of the hemispherical belonging to the course whose template you are trying to construct, drawing the *visual lines* marked with C, until you reach the perpendicular line marked with D" (our italics).[30] The connection of this term with perspective seems even clearer in Ginés Martínez de Aranda [c. 1600, 67]. He uses the cone-development method to construct the intrados templates of a peculiar arch with a spherical intrados, the *Arco en vuelta de horno por la cara*; when talking about the generatrix of the cone he lacks a word for this line and dubs it again as "visual line". Since the axis of this cone is horizontal, as in the "Bóveda de Murcia", it is easy to understand why the generatrix is associated to one of Alberti's rays, while the intersection point with the axis, that is, the vertex of the cone, corresponds of course to the station point.

An interesting variation of this scheme is the *Bóveda de Murcia por cruceros*, [Vandelvira c. 1580: 70v] that is, a coffered torus vault in which the coffering is built as a network of ribs, adapting the mainstream Gothic construction method to classical coffering. This scheme, which as far as we know was never actually built, shows a striking resemblance to the hollow rings depicted in Leonardo's *Codex Atlanticus* or the treatises of Barbaro and Lorenzo Sirigati [Esteve 2007]. In any case, there is a closer source for Vandelvira's coffering, a number of works by Diego de Siloé such as the vaults over the crossing and the presbytery in San Jerónimo in Granada and the tunnels between the presbytery and the ambulatory in the Cathedral of Granada.

Compared to the explanation of the steretomical procedure for the Murcia vault, Vandelvira's references to the annular vault in Granada are somewhat disappointing. First, he quotes it when explaining the *Vis de Saint-Gilles*, a staircase covered by an annular ascending vault, in these terms:

> This method is also useful to construct a vault around a circular courtyard,
> as it is done in the Royal Alcázar in the Alhambra in Granada; however, ...
> you should take from the second end the same portion of the block that
> you take from the first end, since this vault lies all at the same level
> [Vandelvira, c. 1580: 53r].[31]

That is, an annular vault such as the one in Granada can be built using a simplification of the procedure proposed by Vandelvira for the *Vis de Saint-Gilles*, taking into account that the vault in the palace of Charles V is not sloping.

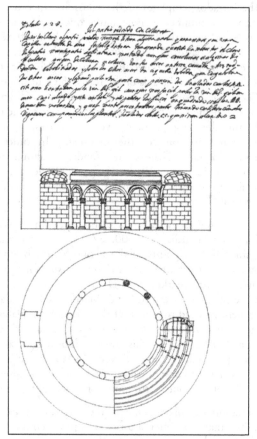

Fig. 21. Round courtyard with columns. Alonso de Vandelvira,
Libro de trazas de cortes de piedras, c. 1580

Later on, Vandelvira includes a section specifically devoted to the problem of vertical-axis torus vaults, under the heading "Patio redondo con columnas", including a drawing of a round courtyard with a perimetric corridor [Vandelvira, c. 1580, 111r] Although the sketch (fig. 21) resembles the central space in the palace of Charles V, once again there are a number of important differences: the columns are joined by arches, not lintels, and the cross-section of the vault is an oval, not a segmental arch. The written explanation is quite short. After stressing the stereotomic difficulties of the warped arches, Vandelvira deals with the vault in two lines, stating that:

> This vault should be understood by any one who knows the *Vía de San Gil*, however, it is easier, since the *Vía de San Gil* is a sloping vault and this one is set at the same level; so the stones should be set square [...] and you should take the same portion of stone from both ends taking into account the bonding as the templates [...] show [Vandelvira, c. 1580: 111r].[32]

Thus we should turn our attention to Vandelvira's explanation of the *Vía de San Gil* [Vandelvira, c. 1580: 52v-53 r], focusing on the basic tracing technique, and leaving aside the steps in the tracing and carving process that are connected with the ascending movement of this spiral staircase. The mason is to trace the plan of the courtyard, including the outside wall and the column line; then he should construct a cross-section of the vault, connected to the plan by means of reference lines, and divide it into voussoirs; once this is done, he should enclose each voussoir in a rectangle with horizontal and vertical sides.

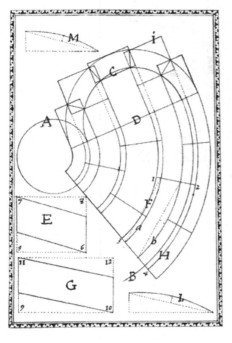
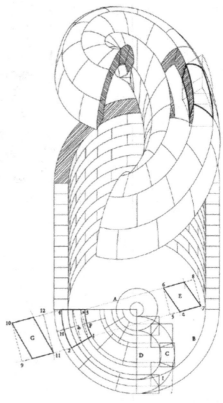

Fig. 22. Barrel-vaulted spiral staircase. Ginés Martínez de Aranda, *Cerramientos y trazas de montea*, c. 1580

Fig. 23. Barrel-vaulted spiral staircase according to Martínez de Aranda. Stonecutting diagram by José Calvo-López

These enclosing rectangles are the trademark of the squaring method for voussoir dressing; rather than relying on templates, as in the "Bóveda de Murcia", this method relies heavily on the orthographic projections of voussoir faces, materialised at the carving phase by means of the square. However, when explaining this step, Vandelvira focuses on the ascending movement of the vault, and does not give clear directions for the essential dressing process. Fortunately, Ginés Martínez de Aranda is a bit more explicit when dealing with the *Vía de San Gil* [1600: 231-233]. Following his directions (fig. 22, 23), the mason should inscribe a wedge shape in the upper face of the voussoir and remove the material outside the wedge by means of the square; doing so, he is in fact reversing the orthogonal projection process and materialising the projecting planes of all voussoir edges with the square. Once this is done, he is to place at both sides of the voussoir an auxiliary template with the portion of the cross-section of the vault that corresponds to the voussoir, to trace these templates and to remove the material outside the shape of the template. That is, he should take out four wedges, corresponding to the intrados, the extrados and both bed joints, controlling the execution of the edges between voussoir faces with the help of a *cerce*, or curved-edge ruler; this process should give as a result the finished voussoir.

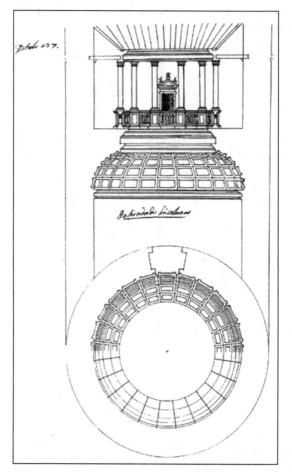

Fig. 24. Round courtyard without columns. Alonso de Vandelvira,
Libro de trazas de cortes de piedras, c. 1580

When discussing the *Vis de Saint-Gilles*, Vandelvira remarks that the squaring method is quite difficult; in a well-known passage, Philibert de L'Orme stresses that it causes "*gran perte de pierres*" [Vandelvira, c. 1580: 52v; De L'Orme 1567: 73v]. In fact, Vandelvira hints that the template method can be used in a level annular vault, although he does so quite indirectly. Immediately before the *Patio redondo con columnas* he includes another design, the *Patio redondo sin columnas* (fig. 24), featuring again a torus vault, cantilevered from the external wall of a round courtyard. Although the intrados surface of the vault features coffering, it is not executed as an independent network of ribs, as in the "Bóveda de Murcia por cruceros", but rather as a superficial decoration, independent from the voussoir-division scheme, as in the Orihuela niches [Vandelvira, c.

1580: 110v]. The *Patio redondo sin columnas* does not include the usual written explanation, but Vandelvira makes a passing reference to this design in the next section, the *Patio redondo con columnas*: "Since the preceding courtyard is quite clear, I have not included any writing, for it is no more than a hemispherical vault and you should leave a lantern as wide as the opening in the courtyard demands..." [Vandelvira, c. 1580: 111r].

In fact, the vault is not a portion of a hemispherical vault, but clearly a part of a torus surface. Thus, we should understand that the reference to the dome applies mainly to the stereotomic technique, and that Vandelvira is suggesting that the cone development method he used in the hemispherical vault can be used in the same way in the *Patio redondo sin columnas*.[33]

Orthographic projections in perspective and stereotomy

As we have seen, Vandelvira uses the squaring method, which makes heavy use of orthographic projections, for the most difficult stonecutting traits, such as those involving warped surfaces [c. 1580: 26r, 46r, 52v, 58 v, 60r]. In the same way, Piero used his "other method", involving direct projection from plan and profile, for the "most difficult shapes" [Della Francesca c. 1480: 32r; see also Di Teodoro 2001; Di Teodoro 2002]. However, the general structure of both *De prospectiva pingendi* and the *Libro de trazas de cortes de piedras* strongly suggests that a painter or a mason dealing in practice with moderately complex problems, such as columns, pedestals, wells, trumpet squinches, skew arches or hemispherical vaults, should not use methods that rely heavily on orthographic projection, but rather centric points, Piero's antecedent of the distance vanishing point construction, rotations or developed templates. These particular methods are used where adequate, but cannot be considered general methods, since they cannot solve all problems; for example, the centric point is useless when trying to construct the perspective of an oblique column. In the same way, Vandelvira cannot use true shape templates in the *Vis de Saint-Gilles* or the groin vault, taking into account the complex nature of these surfaces, and needs to resort to orthographic projections and the squaring method [Vandelvira, c. 1580: 25r, 52v].[34] Thus, both in perspective and stereotomy, orthographic projections furnish a general operative method, although in both fields, a number of special purpose methods furnish more efficient solutions to specific problems, such as the centric point, the distance point, vanishing points, rotations, rabattements and developments.

However, this is not the only use of orthogonal projections in stereotomy and perspective. As we have said, Vandelvira gives two different solutions for the *Arco en torre redonda y cavada*, an arch opened in a curved wall [Vandelvira, c. 1580: 22 r, 24v]. When explaining the first solution, using full-scale templates, he fears that the result will befuddle the reader. To leave any doubt aside, he advises the reader to make a model of the arch by squaring: "if you want to prove it, made a model by squaring, as I will teach you further on, and apply the templates to the model, and you will find that the templates are correct" [Vandelvira, c. 1580: 22r]. Four pages later, he explains the procedure to dress the voussoirs of the same arch by means of squaring, almost excusing himself for using such stone-consuming method: "I will show now the arch in a round wall by squaring, because I had promised to do so, and also to cast light on other pieces that can be done only by squaring". Thus, if Vandelvira applies the squaring method to this arch, it is because he grants it stronger empirical evidence, brought off by means of models [Vandelvira, c. 1580: 24 v].[35]

Empirical evidence plays a central part in the historical developments in perspective, although it was usually furnished by other means. As Filippo Camerota [2001] has suggested, most probably the strange profile of the upper edge of Brunelleschi's panel of the Piazza della Signoria allowed him to superimpose the profile over the Signoria skyline, verifying in a most empirical way the exactitude of his perspectival constructions.

As perspective methods were developed, a number of physical devices provided this empirical evidence, serving at the same time as operative methods, such as Alberti's *velo*, Dürer's and Keser's frames, nails and ropes, Vignola-Danti machines or the *camera obscura* [Alberti 1435, II-31; Dürer 1525, 90v, 91r; Vignola 1682:56;[36] Kemp 1990: 167-203]. However, orthogonal projections stand between empirical evidence and the ingenious centric point method. Following Alberti's trail, Piero gave a proof of the role of the centric point as the vanishing point of the lines that are orthogonal to the picture plane, and did so using double orthogonal projection [Della Francesca c. 1480: 6r-6v; Elkins 1987; Field 2005: 141-146].[37] In a similar way, in Vandelvira's manuscript, orthogonal projections stand between the purely empirical evidence furnished by reduced-scale models and the more elaborate template method.

Thus, Pérouse de Montclos was following a sound trail: both in perspective and stereotomy, orthographic projection plays the dual role of a general operative method and a connection between the empirical basis of both fields and their particular specific-purpose methods. Perspective and stereotomy have also in common a number of techniques based on orthogonal projections, such as the use of rotations and changes of projection plane, which were much later to be accepted between the methods of descriptive geometry, in the strict sense of the term.

However, we must not jump to conclusions and surmise that the geometrical knowledge of Spanish Renaissance masons derives from perspective treatises. Alberti's works on perspective were not translated into Spanish until the late eighteenth century, while Piero still awaits his turn [Rejón de Silva 1784]. The main sources on perspective available to Spanish Renaissance masons were Serlio's second book, not translated into Spanish but enjoying a certain circulation, and Pedro Ambrosio de Ondériz translation of Euclid's *Optics*, dated 1585, while Vandelvira was writing his manuscript [Serlio 1545; Euclid 1585]. Hernán Ruiz includes both perspective and stonecutting problems in his manuscript [c. 1550: see for instance 46v-47v, 51r-58v] but some of his perspectival methods are not connected with the tradition of Brunelleschi, Alberti and Piero, since the orthogonals do not always meet at a single centric point, as Gentil [1998] has stressed. There are references to "visual lines" in Alonso de Guardia [c. 1600: 87v] and Ginés Martínez de Aranda's [c. 1600: 67] manuscripts, which seem connected to Alberti's pyramid, although they could also derive from Euclid's *Optics*, by way of the translation by Ondériz, and this is about all. Of course, a number of problems and methods could have been introduced into Spain by Italian or Italianate artists such as Torni, Machuca or Siloé, in particular the use of torus vaults, but this can hardly account for the use of cone developments in Seville at such an early date as 1543 [Ruiz de la Rosa: 2002].

Thus, Quattrocento perspective and Spanish sixteenth-century stereotomy share a number of concepts, problems and methods; however, there seems to be no direct substantial connection between them. This suggests the existence of a common source, but it is not easy to identify it. Neither classical geometry [Euclid c. -250: books XI-XIII] nor the mediaeval tradition of practical geometry, spanning from Hugh of Saint Victor [c. 1125] to Mathes Roriczer's *Geometria Deutsch* [c. 1490] include a word about orthographic projections, rotations or projection planes.[38] In fact, orthographic projection is almost completely lacking from architectural drawings and full-scale tracings in Antiquity, and only appears clearly in the High Gothic period [Sakarovitch 1997: 21-31, 35-51]. Thus, mediaeval construction shop practices furnish the most probable common source for perspectival and stereotomic methods. Curiously, these practices are seldom mentioned in the exhaustive literature on perspective; even the use of orthogonal projection, although impossible to deny, is not often stressed. On the other hand, the Gothic tradition is recognised, at least in Spain, as an important source of Renaissance stereotomic methods, although this debate is still open; by contrast, the role of Italy, Italian and Italianate artists and perspective in this field is almost taboo.

Acknowledgments

This article is part of the research project "Ashlar construction in the Hispanic area. Written sources and built heritage", which is funded by the Ministry of Science and Innovation of Spain (BIA2006-13649). We wish to show our gratitude to the Colegio de Arquitectos de Murcia for their kind permission to use David Frutos' photograph of the Chapel of Junterón, to María Dolores López for her help and to Richard Etlin for his kind remarks.

Notes

1. It can be argued that cylindrical projections can be understood as a particular case of conical projections, with the vertex of the cone placed at infinity. However, when seventeenth-century treatises, such as Jean-Charles de la Faille's *Tratado de la arquitectura* [c. 1640] compare the vertex positions in cylindrical and conical projections, they do so to stress the differences between the two kinds of projection. Although the works of Girard Desargues, also dating from around 1640, hint at a common foundation for cylindrical and conical projections, the idea was fully developed only by nineteenth-century projective geometry. Thus, if we understand the idea of connections between perspective and stereotomy in general terms, it seems rather anachronistic at first sight.
2. The use of stereotomic tracings in frontispices was not new; it had been put into practice by Philibert de L'Orme in *Le premier tome de l'architecture* [1567]. However Goiti gave it a new turn, since both side tracings act as emblems of the two basic stereotomic methods.
3. Field offers a precise and detailed explanation of Piero's perspective techniques ranging from [2005: 129-173], but deals with the torus in a just a few lines [2005: 168].
4. Roccasecca [2001b: 95-96] has pointed out that the construction of the rings in drawings 1756A and 1757A does not follow exactly Piero's method, since in 1756A and 1757A there are compass marks at the center of the meridians and inkless lines joining these centers with the corners of the meridians; also, in 1757A there is an inkless line following a parallel.
5. To be precise, here Piero seems to be using the side of the enclosing square, which equals twice the apothem of the octagon, as an approximation to the lesser diameter of the torus, which should equal the radius of the octagon.
6. Page numbers are taken from the electronic facsimile in Bibliotheca Perspectivae.
7. See in particular the "Postcript" to the English translation in pages 39-41, where he explains that taking into account the 1966 edition of *De re aedificatoria*, the letter to Leo X must be read as an explanation of a passage in Alberti's text explaining orthographic projection, mistranslated in Bartoli's edition.
8. It does not make much sense in this case to classify these rays as centric, median or extrinsic. To start with, the whole perspective falls below the centric point. Besides, when Alberti differentiates between median and extrinsic rays, he is thinking of a figure with a simple contour or, as topologists put it, a simply connected figure. However, the ring is a double connected figure, since it has a hole in the middle. Thus, there are two contours in the perspective, given the particular eye point Piero has chosen, and two sets of extrinsic points, one external and one internal.

9. Each of the 96 points has a different projecting line. However, the torus is symmetric about a horizontal plane and about a good many vertical planes, including a plane that is parallel to the projection plane of the profile. Thus, each point in the plan is superimposed to another point and the same thing happens in the profile; as a result, the draftsman should trace 48 lines in plan as well as 48 lines in the profile. However, he must transfer each of these intersections twice to the perspective, since each intersection represents two different points; this amounts to 192 transfers. It is not surprising, then, that Piero advises the reader to use a large set of rulers to avoid being misled by such a multitude of points.

10. Garcia-Salgado's terminology, although unusual, is useful here, since a number of studies (for example [Panofsky 1927; Klein 1961; Field 2005]) use the term "distance point" in the sense of "distance vanishing point" while others claiming the use of "distance point" in Alberti or Piero must mean the "distance to the perspective plane".

11. There is an interesting exception. In fact, in fol. 11 r., when solving a rather far-fetched problem – how to cut a square from a rectangle of unknown proportion –, Piero gives sound instructions on how to use the "distance vanishing point" in order to construct a square lying on the ground plane; he does so by placing the "distance vanishing point" on the horizon line at the correct distance and tracing a diagonal through the "distance vanishing point", without using projection and intersection in the profile. However, when he tries to present the proof of this construction, his demonstration is unsound. It is quite remarkable that Piero does not use this construction again; thus, he seems to grasp the notion of "distance vanishing point" on an empirical basis, but he is not sure enough of its foundation to use it as an operative method and in fact this passage is lacking from the Latin version of the manuscript; see [Field 2005: 148-150].

12. In fact, Uffizi 832 Ar includes a plan of the ring that could have been used for this purpose, while Richard Talbot [2006], working independently from Roccasecca, has posited such a solution for the *Chalice*. It is worthwhile to remark that similar methods are suggested by Alberti [1435: II-34], present in Serlio [1545: 33r, 34r, 35r] and ubiquitous in Cousin [10r, 21r, 22v, 26r, etc.]. However, this technique does not solve all the problems involved in the construction of the rings or the chalice. Once the draftsman has constructed the intersection of a meridian plane and the ground plane, he must raise it to its real horizontal plane; this problem can be solved using the method explained by Piero in his second book [Della Francesca c. 1480: 18r-24v; Field 2005: 156-159]. After this, the artist must place each corner of the ring in the corresponding line. Roccasecca has remarked that Uffizi 1757A includes an inscribed elliptical line connecting all the corners at the middle level of the ring, while Uffizi 830Ar, labelled "*Per fare uno mazzocchio*" includes Dürer's [1525: 16r] well-known construction for a conical section. Roccasecca [1998] suggests that the draftsmen of these toruses could have used a number of ellipses to place the corners of the ring along the horizontal lines, constructed maybe using an ellipsograph; in fact, 832Ar includes a singular ellipsograph as well as a scheme for the gardener's ellipse. In any case, there is a final problem to be solved, which is not mentioned by Roccasecca: the center of the resulting ellipse is not the projection of the center of the original circle. In our case, the draftsman can overcome easily this difficulty, since he can place both ends of the shorter axis of the ellipse and divide the resulting segment in two equal parts to place the center of the ellipse. However, it is not so easy to find the ends of the longer axis of the ellipse, in order to use the ellipsograph, since they do not correspond to a diameter of the original circle; at this point the draftsman of 1757A probably used some technique based on trial and error.

13. Della Francesca [c. 1480: 37v]: *Poi piglia il sexto et circula la quantità che tu intendi fare grande il torculo, et il suo centro sia M, et il circulo sia circulo A; poi tira MA linea recta, er con lo sexto piglia la quantità de FB del quadrecto e polla su la linea MA principiando da A, et dove termina l'altro pie del sexto su la dicta linea AM segna B; poi piglia il sexto e poni un pie sopra M et l'altro stendi perfine ad B et descrivi il circulo, che sia il circulo B*, and so on until all four circles are constructed.

14. Della Francesca [c. 1480: 37v]: *benché sieno proposti octo circuli, in questa demostratione faremo con quactro, perchè poremo il dicto torculo giacente piano. Ma quando giacesi altrimente, siria necessario che fussino tante righe quanti circuli (e) in esso torculo se contene.*

---

15. An exception is the ring in the Gubbio studiolo; see [Raggio 1996: 23].
16. Of course, all these rings could have been drawn by empirical means, using a *velo* to depict a wooden model, or preparing a perspective of a flat-lying torus on a sheet of paper, applying it to the painting and rotating it until it fitted over the head of one of the fighters in San Romano or around the neck of a character in the *Flood*. However, we hope that the next paragraphs will make clear that these explanations are unlikely, since Piero knew the problem and hints at a different solution.
17. *In questa demostratione faremo con quactro, perchè poremo il dicto torculo giacente piano. Ma quando giacesse altramente, siria necessario che fussino tante righe quanti circuli (e) in esso torculo si contene* [Della Francesca c. 1480: 37v].
18. Incidentally, these bases, which appear in the *Flagellation* in Urbino and in the *Annunciation* and the *Meeting of Solomon and the Queen of Sheba* in San Francesco in Arezzo are the only instances of toruses or polyhedral rings in the extant paintings of Piero.
19. In order to do so, he places the base line of the cube at a sloping angle at will; then he traces perpendiculars to these sloping lines to represent the edges of the cube that were originally in a vertical position. These lines are now sloping, although they are still parallel to the vertical projection plane, and thus the draftsman can easily measure their lengths. Since the rotation axis is perpendicular to the vertical projection plane, the plane of rotation will be parallel to this plane; as a result, the next face will present no difficulty, since the horizontal projection of each point will move along a line that is perpendicular to the reference lines. Tracing these lines and finding their intersections with the corresponding reference lines, Piero can easily place each point in the plan of the rotated cube.
20. Page numbers are taken from the electronic facsimile in Bibliotheca Perspectivae.
21. Although the title page of the 1545 edition of Serlio is *Il primo libro di architettura, di Sebastiano Serlio, Bolognese*, this volume includes both Serlio's first book on Geometry and his second book on Perspective, and of course the explanation of the perspective of the groin vault belongs to the second book.
22. Leaving aside the tiresome procedure needed to construct the ellipse, which was mastered by Vandelvira but probably unknown to Gothic masters, this rotation furnishes a simple method for showing the diagonal rib and the tierceron in true shape and to control the curvature of these ribs, which is essential in the dressing of springers, keystones and ordinary voussoirs; an elliptical projection would have been useless for this purpose.
23. The technique used by Philibert in the *Nouvelles inventions* [1561] is quite interesting, since he constructs an oblique elevation of a groin starting from a standard elevation, but has no direct connection to our subject.
24. Íñigo López de Mendoza was a key figure in the first Spanish Renaissance. He built the monastery of San Antonio in Mondéjar as early as 1489 and supported his uncle, Pedro González de Mendoza, in his striking decision to build a Renaissance façade in the college of Santa Cruz in Valladolid in 1487. Paradoxically, he was in charge of the supervision of the Gothic building of the Royal Chapel in Granada, although he died before the arrival of Machuca and Torni [Gómez Moreno:1925a; 1925b]
25. Literally, "*está puesto por obra en una villa que se llama San Gil en Francia*" [Vandelvira c. 1580, 52v].
26. Another line, the shorter proportion of the central rectangle in comparison with the Murcia vault, is not so easy to justify, and seems to be related simply to the width of the sheet of paper Vandelvira is using. This small detail is not negligible: as a result, the intrados surface is not a torus from the topological point of view; in fact, if the inner half of the torus were to be materialised, the surface would intersect itself. However, this has no consequences in the actual building process.
27. This meridians are half circles; however, they are depicted as straight lines in the elevation since they lie on plane that is perpendicular to the projection plane of the elevation.
28. This procedure raises two interesting questions, although neither of them can be analysed here in full detail. First, why does Vandelvira bother to construct these ellipses? It is quite usual in stonecutting tracings to leave out any line that is not essential for the ultimate purpose of the tracing: to construct the templates of the faces of the voussoirs or to compute the angles

between voussoir edges. However, the meridians of the *Bóveda de Murcia* are not necessary for the construction of the templates and Vandelvira does not mention the angles between voussoir edges or *saltarreglas*. Quite probably, he tackled this complex problem because he intended to lay out the tracing exactly below the vault to be built and to control the position of the voussoirs by means of a plumb line once they were set in place; in fact, he refers twice to this technique [Vandelvira c. 1580: 23 r., 23 v]. Second, what is the nature of this curve? The horizontal projections of the meridians are half-ellipses, but the resulting curve is not, since it is an assembly of arcs of a circle. However, if the purpose of these curves is just to control the placement of the voussoirs, it is easy to surmise that in a real stonecutting tracing the masons would not care to join the points, and that in the *Libro de trazas* ... the curve is drawn in order to show the shape of the meridians for the benefit of the reader.

29. The reason for this striking omission can be found in Vandelvira's explanation of the hemispherical vault, or as he puts it, the "beginning and example of all Roman vaulting" [Vandelvira c. 1580, 60v]. Once he has drawn the arcs that stand for the parallels, he advises the reader that "*las cuales dos cerchas cerrarás por do quisieres*"; that is, "you can close these arcs as you will". Such a cavalier attitude is quite understandable if we take into account that in sixteenth century the stone did not usually arrive from the quarry to the construction site in standard sizes; thus, playing with the length of the voussoir allowed the mason to use the stones coming from the quarry in the most economical way. In fact, this technique was used in such a carefully executed work as the dome over the crossing of the Escorial basilica, where the voussoirs of each course show different lengths. However, this approach cannot be used in a horizontal-axis torus vault, since the courses are not placed between two parallels, but rather between two meridians. Thus, the builders of the vault in Junterón's chapel must have resorted to some device for controlling the length of the voussoirs, although Vandelvira does not say a word about this issue.

30. Guardia [c. 1600: 87v]: "*y para sacar las plantas por caras de las hiladas desta dicha pechina pondras la Regla en la buelta de horno en los dos puntos de la hilada que quisieres sacar y tiraras las linias bisuales señadas con la C que bayan a parar a la linia perpendicular señalada con la D*".

31. Vandelvira [c. 1580: 53 r]: "*Sirve también esta traza para hacer una bóveda alrededor de un patio redondo como está puesto por obra en la alcázar real del Alhambra de Granada, mas antes de trazar las piedras igualmente por entrambos cabos que lo que se roba por la una parte se ha de robar por la otra, por andar a nivel y no capialzar más la una parte que la otra, y así las piezas no han de ser más altas que la tardosa del arco a las líneas de puntos*". Another interesting problem is the source of Alonso de Vandelvira's knowledge of the vault in the palace. According to Galera [2000: 27-28] Andrés de Vandelvira could have seen the model of the courtyard, prepared in 1532. In Galera's opinion, this would explain the differences between the drawing in the *Libro de trazas* and the built courtyard, such as the different height of the springing of the vault in both sides. However, we have seen that Alonso made similar changes in the *Bóveda de Murcia*; this is quite natural, since he was trying to furnish general models for horizontal or vertical torus vaults, and not surveys of the Murcia and Granada vaults; thus, it is easier to assume that Alonso himself could have seen the completed lower gallery, since he describes other works in Granada, such as the Chancillería staircase, completed in 1578 after his father's death.

32. Vandelvira [c. 1580: 52v-53r]: "*Este arco se entenderá por la vía de San Gil aunque más fácil, por ser la vía de San Gil un arco capialzado y éste a nivel y así puestas las piezas en cuadrado como las B.B. demuestran, robarse han igualmente por entrambas partes teniendo consideración a sus ligazones como parecen en las plantas señaladas con las C.C. que miran al centro.*"

33. In fact, Vandelvira [c. 1580: 69v, 103v] also puts on equal standing the *Bóveda de Murcia*, a torus vault and the *Ochavo de la Guardia*, a coffered quarter-sphere vault.

34. We are quoting here in reference to the original folio numbers, given by Barbé in round brackets.

35. Vandelvira [c. 1580: 24v]: "*Porque dije de enseñar el arco torre cavado y redondo por robos y porque también sea lumbre para entender otras trazas que no se pueden hacer si no es por*

robos, pongo ahora éste... ". Again, the quotation is referenced to the original folio numbers of the manuscript, given by Barbé in round brackets.

36. Page numbers are taken from the electronic facsimile in Bibliotheca Perspectivae.
37. Elkins states that the proof "relies almost exclusively on a two-dimensional system of proportions" [1987: 220] but this is misleading, since the visual rays, picture plane and perspective projections are not lying in the same plane and Piero uses orthogonal projection to represent them; in fact, later on Elkins mentions side and top views [1987: 224].
38. Of course, Roriczer's *Fialenbuchlein* [1486] uses orthographic projections; in fact, it is the first text to explain the transfer of measures from plan to elevation, an antecedent of the reference lines used in descriptive geometry. But this is precisely the point: the *Fialenbuchlein*, deriving from workshop practice and oral traditions, uses orthographic projection, while the *Geometria Deutsch* [1490 c.], which derives from *De inquisitione capacitatis figurarum*, as Shelby has shown [1972: 412-416], does not mention projections at all.

References

ALBERTI, Leon Battista. 1435. *De Pictura*. [Latin Ms., translated by the author to Tuscan in 1436].
———. 1966. *De Re Aedificatoria*. Milano: Il Polifilo. (Ed. with latin text and translation by Giovanni Orlandi. Original manuscript c. 1440. 1st. ed. 1485. Firenze: Nicolai Laurentii Alamani).

BARBARO, Daniel. 1569. *La pratica della perspettiva : opera molto vtile a pittori, a scvltori, & ad architetti: con due tauole, una de' capitoli principali, l'altra delle cose piu notabili contenute nella presente opera*. Venice: Camillo & Rutilio Borgominieri.

BARBE-COQUELIN DE LISLE, Geneviève. 1990. La pierre et l'architecture de la Renaissance espagnole: un matériau symbolique. *Boletín del Museo e Instituto 'Camón Aznar'* (42): 31-38.

BECHMANN, Roland. 1993. *Villard de Honnecourt. La penseé technique au XIIIe siécle et sa communication*. Paris: Picard. (1st ed. 1991).

CALVO CASTELLÓN, Antonio. 1994. Pinturas italianas y españolas. Pp. 214-229 in *El libro de la Capilla Real*. Granada: Miguel Sánchez.

CALVO-LÓPEZ, José. 1999. 'Cerramientos y trazas de montea' de Ginés Martínez de Aranda. Ph.D. dissertation, Universidad Politécnica de Madrid. http://repositorio.bib.upct.es/.
———. 2003. Orthographic projection and true size in Spanish stonecutting manuscripts. Pp. 461-471 in *Proceedings of the First International Congress on Construction History*. Madrid: Instituto Juan de Herrera.
———. 2005a. Jacopo Torni l'indaco vecchio and the emergence of Spanish classical stereotomy. Pp. 505-516 in *Teoria e Pratica del costruire: saperi, strumenti, modeli*. Bologna-Ravenna: Università di Bologna-Fondazione Flaminia.

CALVO-LÓPEZ, José, and Miguel Ángel ALONSO-RODRÍGUEZ. 2005b. Bóvedas renacentistas de intradós esférico y tórico en el antiguo obispado de Cartagena. Pp. 67-84 in *XVI Jornadas de Patrimonio Histórico. Intervenciones en el patrimonio arquitectónico, arqueológico y etnográfico de la Región de Murcia*. Murcia: Consejería de Cultura.

CALVO-LÓPEZ, José, Miguel Ángel ALONSO-RODRÍGUEZ, Enrique RABASA DÍAZ, and Ana LÓPEZ MOZO. 2005c. *Cantería renacentista en la catedral de Murcia*. Murcia: Colegio de Arquitectos

CALVO-LÓPEZ, José, Miguel Ángel ALONSO-RODRÍGUEZ and María del Carmen MARTÍNEZ RÍOS. 2008. Levantamiento y análisis constructivo de la cabecera de la iglesia de Santiago de Jumilla. Pp. 649-659 in *XIX Jornadas de Patrimonio Cultural*. Murcia: Consejería de Cultura.

CAMEROTA, Filippo. 2001. L'esperienza di Brunelleschi. Pp. 27-31 in *Nel segno di Masaccio. L'invenzione della prospettiva*. Firenze: Giunti.

COUSIN, Jean. 1560. *Livre de prespectiue de Iehan Cousin*. Paris: Iehan le Royer.

DAVIS, Margaret Daly. 1980. Carpaccio and the perspective of regular bodies. Pp. 123-132 in *La prospettiva rinascimentale. Codificazioni e trasgressioni*. Milano: Centro Di.
———. 2001. Il disegno dei corpi regolari. Pp. 123-127 in *Nel segno di Masaccio. L'invenzione della prospettiva*. Firenze: Giunti.

DEL MONTE, Guidobaldo. 1600. *Gvidivbaldi è marchionibvs Montis Perspectivae libri sex*. Pesaro: Hieronymum Concordiam.

DE LA FAILLE, Jean-Charles. 1640 c. *De la arquitectura*. Ms. in Biblioteca del Palacio Real, Madrid.

DELLA FRANCESCA, Piero. 1480 c. *De prospectiva pingendi*. (Ed. G. Nicco-Fasola, 1980, Firenze: Le Lettere).

DI TEODORO, Francesco Paolo. 2001. Piero della Francesca, il disegno, il disegno d'architettura. Pp. 112-116 in *Nel segno di Masaccio. L'invenzione della prospettiva*. Firenze: Giunti.

————. 2002. Vitruvio, Piero della Francesca, Raffaello: note sulla teoria del disegno di architettura nell' Rinascimento. *Annali di Architectura* (14):35-54.

————. 2003. *Rafaello, Baldassare Castiglione e la Lettera a Leone X.* San Giorgio di Piano: Minerva Edizioni.

DÍEZ DEL CORRAL Garnica, Rosario. 1992. Arquitectura y magnificencia en la España de los Reyes Católicos. Pp. 55-78 in *Reyes y Mecenas. Los Reyes Católicos-Maximiliano I y los inicios de la casa de Austria en España*. Madrid: Ministerio de Cultura.

DÜRER, Albrecht. 1525. *Underweysung der messung mit dem zirkel und richtscheyt ....* Nuremberg. [Electronic facsimile of the 1538 ed. in Bibliotheca Perspectivae, www.imss.fi.it].

————. 1532. *Alberti Dureris ... de symmetria partium in rectis formis humanorum corporum*. Nuremberg. (Latin translation of *Vier bucher von menslicher proportion*. 1532; electronic facsimile in Bibliotheca Perspectivae, www.imss.fi.it.)

ELKINS, James. 1987. Piero della Francesca and the Renaissance Proof of Linear Perspective. *The Art Bulletin* **69**, 2: 220-230

ESTEVE SECALL, Carlos. 2007. El mazzocchio: paradigma de investigación geométrica en el Renacimiento. *Revista de Expresión Gráfica Arquitectónica* (12):106-113.

ETLIN, Richard. 2009. Génesis y estructura de las bóvedas de Arles. In *VI Congreso Nacional de Historia de la Construcción*, forthcoming.

EUCLID. 1585. *La perspectiua y especularia de Euclides. Traduzidas en vulgar Castellano ... por Pedro Ambrosio Onderiz ...* Madrid: Viuda de Alonso Gómez.

EVANS, Robin. 1995. *The Projective Cast*. Cambridge, MA: MIT Press.

FIELD, Judith Veronica. 2005. *Piero Della Francesca. A mathematician's art*. New Haven: Yale University Press.

FRANKL, Paul, and Erwin PANOFSKY. 1945. The secret of the mediaeval masons: with an explanation of Stornaloco's formula. *Art bulletin* **27**: 46-60.

GALERA ANDREU, Pedro. 2000. *Andrés de Vandelvira*. Madrid: Akal.

GARCÍA-SALGADO, Tomás. 2003. Distance to the perspective plane. *Nexus Network Journal* **5**, 1: 22-48.

GELABERT, Joseph. 1977. *De l'art de picapedrer*. Palma de Mallorca: Instituto de Estudios Baleáricos. (Facsimile ed. of a 1653 ms. in the library of the Consell Insular de Mallorca, Palma de Mallorca.)

GENTIL BALDRICH, José María. 1998. El libro de perspectiva. Pp. 215-234 in Hernán Ruiz el Joven. *Libro de arquitectura*. Sevilla: Fundación Sevillana de Electricidad.

GÓMEZ-MORENO, Manuel. 1925a. Sobre el Renacimiento en Castilla. Notas para un discurso preliminar. *Archivo Español de Arte y Arqueología:* 1-40.

————. 1925b. En la Capilla Real de Granada. *Archivo Español de Arte y Arqueología*: 225-288.

GONZÁLEZ SIMANCAS, Manuel. 1905-1907. Catálogo Monumental de España. Provincia de Murcia. [Facsimile ed. 1997. Murcia: Colegio de Arquitectos].

GUARDIA, Alonso de. 1600 c. *Manuscrito de arquitectura y cantería*. Notes and drawings on a copy of Battista Pittoni, *Imprese di diversi principi, duchi, signori ...*, Book II, Venezia, 1566. Madrid, Biblioteca Nacional, ER/4196.

GUTIÉRREZ-CORTINES CORRAL, Cristina. 1987. *Renacimiento y Arquitectura religiosa en la antigua diócesis de Cartagena*. Murcia: Colegio de Aparejadores y Arquitectos Técnicos.

KEMP, Martin. 1990. *The science of Art*. New Haven: Yale University Press.

KERN, G[uido] I[osef]. 1915. Der Mazzocchio des Paolo Uccello. *Jahrbuch der preussichen Kunstsammlugen* 36: 13-38.

KLEIN, Robert. 1961. Pomponius Gauricus on Perspective. *The Art Bulletin* **43**, 3: 211-230.

LA GOURNERIE, Jules-Antoine-René Maillard de. 1860. *Traité de géométrie descriptive, par Jules de La Gournerie,...* Paris: Mallet-Bachelier.

LÓPEZ DE TORO, José. 1953. *Documentos inéditos para la Historia de España. Tomo IX. Epistolario de Pedro Mártir de Anglería*. Madrid.

———. 1955. *Documentos inéditos para la Historia de España. Tomo X. Epistolario de Pedro Mártir de Anglería*. Madrid.

L'ORME, Philibert de. 1561. *Nouvelles inventions pour bien bastir a petits frais*. Paris: Federic Morel.

———. 1567. *Le premier tome de l'Architecture*. Paris: Federic Morel.

LOTZ, Wolfgang. 1977. The rendering of the interior in architectural drawings of the Renaissance. Pp. 1-41 in *Studies on Italian Rennaissance Architecture*. Cambridge, Mass., MIT, 1977. (Originally published as Das Raumbild in der architekturzeichnung der italianischen Renaissance. *Mitteilungen des Kunsthistoriches Instituts in Florenz* **6**: 193-226.)

MARÍAS, Fernando. 1989. *El largo siglo XVI*. Madrid: Taurus.

MARTÍNEZ DE ARANDA, Ginés. 1600 c. Cerramientos y trazas de montea. (Ms. in the library of the Servicio Histórico del Ejército, Madrid. Facsimile ed., Madrid, Servicio Histórico del Ejército-CEHOPU, 1986.)

MESEGUER FERNÁNDEZ, Juan. 1957. Nuevos datos sobre los hermanos Valdés: Alfonso, Juan, Diego y Margarita. *Hispania* **17**, 68: 369-394.

MORALES, Alfredo J. 1992. Italia, los italianos y la introducción del Renacimiento en Andalucía. Pp. 177-197 in *Reyes y Mecenas. Los Reyes Católicos-Maximiliano I y los inicios de la casa de Austria en España*. Madrid: Ministerio de Cultura.

MÜLLER, Werner. 1990. *Grundlagen gotischer Bautechnik. Ars sine scientia nihil est*. München: Deutscher Kunstverlag

PALACIOS GONZALO, José Carlos. 2003. *Trazas y cortes de cantería en el Renacimiento Español*. Madrid: Munilla-Llería. (1st ed., 1990. Madrid: Ministerio de Cultura.)

PANOFSKY, Erwin. 1927. Die Perspektive als 'Symbolische Form'. In: *Vorträge der Bibliothek Warburg*. Berlin: Teubner.

PELERIN, Jean. 1505. *De artificiali perspectiva*. Toul: Petri Iacobi

PERGAEUS, Apollonius. 1566. *Conicorum libri quattuor*. Bologna: Alexandri Benatii. (Translation by Federico Commandino.)

PEROUSE DE MONTCLOS, Jean-Marie. 2001. *L'Architecture a la française*. Paris: Picard [1st ed. 1982].

RABASA DÍAZ, Enrique. 1996. Técnicas góticas y renacentistas en el trazado y la talla de las bóvedas de crucería españolas del siglo XVI. Pp. 423-433 in *Actas del Primer Congreso Nacional de Historia de la Construcción*. Madrid: Instituto Juan de Herrera.

———. 2000. *Forma y construcción en piedra. De la cantería medieval a la estereotomía del siglo XIX*. Madrid: Akal

RAGGIO, Olga. 1996. The Liberal Arts Studiolo from the Ducal Palace at Gubbio. *The Metropolitan Museum of Art Bulletin* **53**, 4: 5-35.

REJÓN DE SILVA, Diego Antonio. 1784. *El tratado de la pintura por Leonardo de Vinci y los tres libros que sobre el mismo arte escribió Leon Bautista Alberti. Traducidos é ilustrado con algunas notas por Don Diego Antonio Rejon de Silva...* Madrid: Imprenta Real.

ROCCASECCA, Pietro. 1998. Tra Paolo Uccello e la cerchia sangallesca: la costruzione prospettica nei disegni di Mazzocchio conservati al Louvre e agli Uffizi. Pp. 133-144 in: *La prospettiva: fondamenti teorici ed esperienze figurative dall'antichità al mondo moderno*. Fiesole: Cadmo.

———. 2000. Il calice degli Uffizi: da Paolo Uccello e Piero della Francesca a Evangelista Torricelli e l'Accademia di Disegno di Firenze. *Ricerche di Storia dell'Arte* **70**: 65-78.

———. 2001a. Paolo Uccello. Pp. 89-92 in *Nel segno di Masaccio. L'invenzione della prospettiva*. Firenze: Giunti.

———. 2001b. Paolo Uccello, Studi di mazzocchi. Pp. 95 in *Nel segno di Masaccio. L'invenzione della prospettiva*. Firenze: Giunti.

———. 2001c. Cerchia dei Sangallo, Modo per disegnare un mazzocchio. Pp. 95 in *Nel segno di Masaccio. L'invenzione della prospettiva*. Firenze: Giunti.

———. 2001d. Studio per un calice. Pp. 118 in *Nel segno di Masaccio. L'invenzione della prospettiva*. Firenze: Giunti.

RORICZER, Mathes. 1486. *Büchlein von der fialen Gerechtigkeit.* Regensburg. (Ed. by Lon R. Shelby, in *Gothic Design Techniques,* Carbondale, Southern Illinois University Press, 1977.)

———. 1490 c. *Geometria Deutsch* (Ed. by Lon R. Shelby, in *Gothic Design Techniques,* Carbondale, Southern Illinois University Press, 1977.)

ROSENTHAL, Earl E. 1985. *The Palace of Charles V in Granada.* Princeton: Princeton University Press

———. 1988. El programa iconográfico-arquitectónico del Palacio de Carlos V en Granada. Pp. 159-177 in *Arquitectura Imperial.* Granada: Universidad de Granada.

RUIZ DE LA ROSA, José Antonio, and Juan Clemente RODRÍGUEZ ESTÉVEZ. 2002. 'Capilla redonda en vuelta redonda' (sic): Aplicación de una propuesta teórica renacentista para la catedral de Sevilla. Pp. 509-516 in *IX Congreso Internacional Expresión Gráfica Arquitectónica. Re-visión: Enfoques en docencia e investigación.* A Coruña: Universidad de A Coruña.

RUIZ EL JOVEN, Hernán. 1550 c. *Libro de Arquitectura.* (Ms. in the library of the School of Architecture of the Universidad Politécnica de Madrid. Facsimile ed., 1988. Sevilla: Fundación Sevillana de Electricidad.)

SAINT VICTOR, Hugh of. 1125 c. *Practica geometriae.* (Ed. by Frederick Homann. 1977. Milwaukee: Marquette University Press.)

SAKAROVITCH, Joël. 1997. *Epures d'architecture.* Basel-Boston-Berlin: Birkhäuser.

SANZIO, Raphael, and Baldassare CASTIGLIONE, attributed to. 1510 c. Letter to Leo X. (Transcribed by Di Teodoro 2003.)

SERLIO, Sebastiano. 1544. *Il libro terzo di Sebastiano Serlio Bolognese nel qual si figurano e descriuono le antiquita di Roma ...* Venezia: Francesco Marcolini.

———. 1545. *Il primo [-secondo] libro d'architettura ... Le premier [-second] livre d'architecture.* Paris: Iehan Barbé [Includes italian text and a French translation by Jean Martin].

———. 1552. *Tercero y quarto libro de Architectura de Sebastian Serlio Boloñes ... Agora nueuamente traducido de Toscano en Romance Castellano por Francisco de Villalpando Architecto.* Toledo: Iuan de Ayala.

SHELBY, Lon R. 1972. The geometrical knowledge of medieval master masons. *Speculum* **47**, 3: 395-421.

———. 1977. Introduction to *Gothic Design Technics: The fifteenth-century design booklets of Mathes Roriczer and Hans Schmuttermayer.* Carbondale: Southern Illinois University Press.

SINDONA, Ennio. 1980. Prospettiva e crisi nell'umanesimo. Pp. 95-124 in *La prospettiva rinascimentale. Codificazioni e trasgression.* Milano: Centro Di.

TAFURI, Manfredo. 1987. Il palazzo di Carlo V a Granada: architettura 'a lo romano' e iconografia imperiale. *Ricerche di Storia dell'arte* (32): 4-26.

TALBOT, Richard. 2006. Design and perspective construction: Why is the Chalice the shape it is? Pp. 121-134 in *Nexus VI: Architecture and Mathematics.* Sylvie Duvernoy and Orietta Pedemonte, eds. Torino: Kim Williams Books.

TOKER, Franklin. 1985. Gothic architecture by remote control: an illustrated building contract of 1340. *Art Bulletin* **67**, 1: 67-95.

TOMLOW, Jos. 2009. On Late-Gothic Vault Geometry. In *Creating shapes in civil and naval architecture. A cross-disciplinary comparison.* Leiden: Brill [forthcoming].

VAGNETTI, Luigi. 1980. Il processo di maturazione di una scienza del arte: la teoria prospettica nel Cinquecento. Pp. 427-474 in *La prospettiva rinascimentale. Codificazioni e trasgressioni.* Milano: Centro Di.

VANDELVIRA, Alonso de. 1580 c. *Libro de trazas de cortes de piedras.* (Ms. in the library of the School of Architecture of the Universidad Politécnica de Madrid. Facsimile ed. by Geneviève Barbé-Coquelin de Lisle, *Tratado de arquitectura de Alonso de Vandelvira.* 1977. Albacete: Caja Provincial de Ahorros.)

———. 1646. *Libro de cortes de cantería de Alonso de Vandeelvira, arquitecto, sacado a la luz y aumentado por Philipe Lázaro de Goiti ...* Madrid, Biblioteca Nacional, Ms. 12.719.

VASARI, Giorgio. 1550. *Le vite de' piú eccelenti architetti, pittori et scultori italiani, da Cimbaue insino a'tempi nostri ...* Firenze: Lorenzo Torrentino.

———. 1568. *Le vite de' piú eccelenti pittori, scultori et architettiori italiani.* Firenze: Giunti.

VELASCO, Lázaro de. 1564. *Traducción de los diez libros de arquitectura de Vitrubio*. Ms. in the State Public Library in Cáceres. Facsimile ed. by Francisco Javier Pizarro Gómez and Pilar Mogollón Cano-Cortés. Cáceres: Cicon, 1999.

VERA BOTÍ, Alfredo. 1993. *La Torre de la Catedral de Murcia. De la teoría a los resultados*. Murcia: Academia Alfonso X el Sabio

VIGNOLA, Giacomo Barozzi da, and Egnazio Danti. 1682. *Le due regole della prospettiva pratica / di Iacomo Barozzi da Vignola; con i comentari del R.P.M. Egnatio Danti dell'ordine de Predicatori ...* Bologna: Gioseffo Longhi. (1st ed. 1583. Roma: Francesco Zanetti.)

VILLELLA, Marzia. 1999. Jacopo Torni detto l'Indaco (1476-1526) e la capella funebre 'a La Antigua' di Don Gil Rodríguez de Junterón nella cattedrale de Murcia. *Annali di Architettura* (10-11):82-102.

———. 2002. Don Gil Rodríguez de Junterón: Comittente Architettonico e Artistico tra Roma e Murcia. *Anuario del Departamento de Teoría e Historia del Arte de la Universidad Autónoma de Madrid* **14**: 81-102.

About the authors

José Calvo-López is an architect. His Ph.D. dissertation was awarded the Extraordinary Doctoral Prize of the Polytechnic University of Madrid in 2001. He is Professor of Graphical Geometry and Academic Director at the School of Architecture and Building Engineering of the Polytechnic University of Cartagena. He has lectured also at graduate courses on stonecutting, stereotomy and the history of spatial representation at Cartagena, at the Polytechnic University of Valencia and at San Pablo-CEU University. His research, focused on stereotomy and other issues concerned with spatial representation, is published regularly on international conferences, journals and such books as *Cantería renacentista en la catedral de Murcia*.

Miguel Ángel Alonso-Rodriguez is an architect and surveying engineer. He holds a Ph. D. in architecture; his dissertation focuses on the history of axonometry. He is Professor of Descriptive Geometry at the School of Architecture of the Polytechnical University of Madrid. He has lectured also at graduate courses on the history of axonometry and surveying at Madrid and the Universidad Nacional Autónoma de Guadalajara. He has carried on surveys of historical buildings or archeological sites in Spain and Pompei both as the result of research grants, such as the ones included in *Cantería renacentista en la catedral de Murcia* or as a commissions from such key Spanish institutions as Patrimonio Nacional or the Real Academia de San Fernando.

Tomás
García-Salgado

National Autonomous University of
Mexico
Palacio de Versalles 200
Col. Lomas Reforma MÉXICO
D. F. C.P. 11930
tgsalgado@perspectivegeometry.com

Keywords: Chichén-Itzá, Kukulcán
Pyramid, light and shadow effect,
Mayan geometry, perspective

Research

The Sunlight Effect of the Kukulcán Pyramid or The History of a Line

Abstract. When the sunlight bathes the Kukulcán Pyramid in the Mayan city of Chichén-Itzá during the equinox sunset, it casts seven triangles of light and shadow that creep downwards along its northeast stairway. According to the *Popol Vuh*, the effect can be interpreted as the myth of the gods of the Heart of Sky coming to the Sovereign Plumed Serpent. Unfortunately, neither the event nor the kind of geometry used to build the pyramid is reported in the extant Mayan codices. There are two reliable facts: first, a line across the pyramid's base coincides with the orientation of the summer-winter solstice; second, an earlier, smaller pyramid is concealed beneath the current one. Hence the major question here is whether the light and shadow effect was intended or occurs accidentally. Perspective as a surveying method for building provides a key to this riddle, because it accounts for a line according to which the Kukulcán Pyramid was built. To tell the history of this line, we have to bring into context other sunlight effects that take place across the Mayan area, taking into account the question of how the buildings in which such effects occur were built.

Historical context of Chichén-Itzá

In the *Chilam Balam* books [2005], according to the Matichu Chronicle, Part II, Chichén-Itzá was discovered[1] during the Katun 6 Ahau (435-455). In 13 Ahau (495-514), the *esteras* (communities) were organized and Chichén-Itzá was occupied. The Itzáes reigned over Chichén-Itzá for two hundred years, but it was abandoned in 8 Ahau (672-692). The Itzáes went to Chakanputún (Champotón), from Katun 6 Ahau (692-711) until 8 Ahau (928-948), when it too was abandoned. Then, they went astray for 40 years (948-987) until they returned to Chichén Itzá. In Part III of the Matichu Chronicle, the account of disputes, wars, and betrayals among the governors of Uxmal, Mayapán, Itzmal, and Chichén-Itzá are briefly described.

Due to these events, in 8 Ahau (1185-1204), the governor of Chichén-Itzá (of the Itzáes) abandoned "their" homes once again. Although the correlation of events is quite confusing in Part III, Chichén-Itzá was probably abandoned in 6 Ahau (1224-1244). Thereafter, in 11 Ahau (1283-1303), the land of the Ichpá-Mayapán was taken by the "men of the city outside the wells" (the Itzáes) and King Ulmil. Several other dates are mentioned until 8 Ahau (1441-1461), when Ichpá-Mayapán was destroyed and definitively abandoned.

A copy of the so-called Maní manuscripts of the *Chilam Balam* was handed to John Lloyd Stephens by Juan Pío Pérez when they met in a small town called Peto. Stephens and Frederick Catherwood arrived in Chichén-Itzá on March 13, 1842 at 4:30 p.m. and left on March 29. In Stephens' own words: "In half an hour we were among the ruins of this ancient city [they departed from Piste at four o'clock], with all the great buildings in full view, casting prodigious shadows over the plain..." [2000: 326]. When the sunlight bathes the Kukulcán Pyramid (fig. 1) in the Mayan city of Chichén-Itzá during the equinox sunset, it casts seven triangles of light and shadow that creep downwards along its northeast stairway. Ironically, they were there at the right time and season to witness the light and shadow effect on the northeast stairway, but they did not see it because the pyramid was largely concealed by the forest. For sixteen days, Stephens explored the ruins while Catherwood drew views of the buildings with the aid of both his camera lucida [2000: 40-41] and daguerreotype apparatus; the latter was destroyed when the horse carrying it ran away during the journey from Chichén-Itzá to Valladolid.

Fig. 1. A charming little Mayan girl passing by the pyramid enhances its majestic presence.
Photograph by the author

Thanks to these instruments, Catherwood's drawings depict many buildings in perspective, stone by stone down to the smallest detail. In one of these drawings, the serpent heads resting at the foot of the stairway of the Kukulcán Pyramid appear quite different from how they look today, an oddity that we will discuss later. Catherwood also drew a plan of the site with the aid of a compass and a line, noticing that none of the buildings coincided with the cardinal points. His plan indicates the road leading to Valladolid passing by the pyramid, and the nearby Hacienda Chichén where he and Stephens stayed (and where my wife and myself stayed in December 2007), just a ten-minute walk from the ruins. The ruins belonged to the Hacienda; in 1894 they were purchased by Edward Thompson, who dredged a *cenote* (a deep natural well) and smuggled many artefacts out of the country.[2] Later, the staff of Carnegie Institution of Washington, led by Sylvanus G. Morley, stayed at the Hacienda Chichén while, over the course of ten years (1924-1934), they explored the ruins, carrying out extensive excavations and mapping the site. Thus, the restoration works began in cooperation with the Mexican government, which in particular undertook the restoration of the Kukulcán Pyramid and the Juego de Pelota (the ball court). This was during the same years when the junior Maya archaeologist Eric S. Thomson wrote his contemptuous opinion about the Observatory, making evident that he was not as good a "connoisseur" of architecture as he said his father was of wine and good food.[3]

Besides sunlight effects, acoustic phenomena also occur in several Mayan cities. In Chichén-Itzá, for instance, standing on the great plaza, around 30 m. away from the center of the pyramid's north stairway, two Quetzal chirps followed by an echo can clearly be heard when you clap your hands once (the Quetzal is the famous long-tailed bird typical of Yucatán). Specialists in acoustics have investigated this effect and believe that the Quetzal's echo could be a sort of incident sound [Declercq et al. 2004]. Another beguiling effect can be experienced when you stand midway in the Juego de Pelota, between its two immense walls, while another person at the end of the field can hear the words you whisper.

Other sunlight effects and building alignments

I would like to bring into context other Mayan cities where significant sunlight effects also take place, to emphasize that those of Chichén-Itzá are not unique. Many buildings across the Mayan area seems to have a line oriented to the summer-winter solstice.

Fig. 2. a, top left) The House of the Seven Dolls in Dzibilchaltun; b, top right) A view of the alignment of the stelas; c, bottom) Schematic plan of the Sacbé 1. Photographs by the author, drawing by Ambar Hernádez

In Dzibilchaltun, 16 km. north of Mérida, capital of the State of Yucatán, when at sunrise, the sun's rays pass through the central door and windows of the House of the Seven Dolls (HSD) during the equinox, an effect is produced resembling a lighthouse (fig. 2a). The HSD is at the extreme east of the sacbé 1 (white road 1), which is oriented 94 az-274 az (east-west),[4] and measures 17 m. wide by 400 m. long. As the sacbé axis and that of the HSD are parallel but not collinear, the sunlight also falls alongside the sacbé, thus separating the path of the men from that of the gods (fig. 2b).

The lighthouse effect also takes place during the solstice sunrise, although less intensely, and depending on the solstice season only one of the windows is illuminated. In turn, on the night of the first full moon following the solstice, moonlight through the central door of the HSD illuminates the sacbé. On the left side of the sacbé's extreme west, there are three stelas perfectly aligned at 190 az (south), coincidentally with the same orientation as the HSD's southern upper window (fig. 2c).

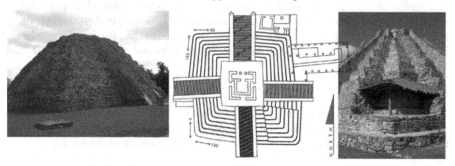

Fig. 3. a, left) The Kukulcán Pyramid in Mayapán; b, center) Notice the lack of symmetry of the plan; c, right) An older substructure is exposed on the southeast corner of the pyramid. Photographs by the author, drawing by Ambar Hernández

In Mayapán, built during the Post-Classical Period (1250-1450 AD), after the fall of Chichén-Itzá, there is another pyramid called Kukulcán, with a light and shadow effect similar to that of Chichén, although it takes place during the winter solstice when the sunlight bathes its southern stairway. This effect is barely perceptible due to the misalignment of the stairways, the inaccurate angles of the pyramid's base (which are off by as much as 10°; its southwest angle points 0-100 az), and the platforms' divergence (figs. 3a and 3b). What we learn from this is that a stepped pyramid produces the effect because of its form; in fact, latitude and orientation can vary while the effect is still produced. Archaeological exploration on the southeast corner of the pyramid shows an older substructure, as in Chichén-Itzá, although here no vault beneath the stairways was left (fig. 3c). Like Chichén-Itzá, Mayapán also had an Observatory, a circular structure whose upper part is missing; it was elevated from the ground to gain view from its four doors, of which the west door is oriented (275 az) toward the northern third platform of the pyramid. It is said that Mayapán is Chichén-Itzá in miniature, but Mayapán was built in a rough way since one can hardly find two parallel lines there.

The architecture of Uxmal, 78 km. south of Mérida, is the most graceful of the Mayan world.

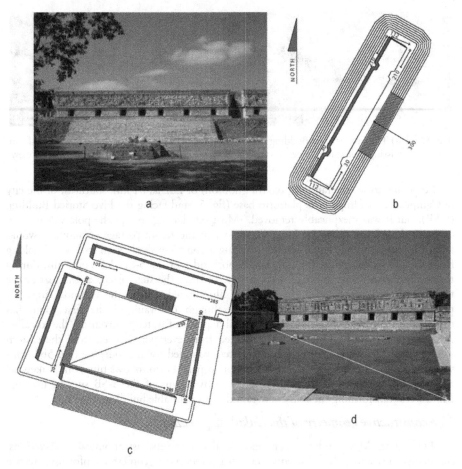

Fig. 4. a, top left) The Governor's palace in Uxmal viewed from the plaza. A broken stone-pole remains on the floor of the small platform at the midst of the plaza, facing the palace; b, top right) Plan of the palace; c, bottom left) Plan of the Nunnery Quadrangle courtyard with its approximated solstice line. d, bottom right) Standing on the courtyard, you cannot tell whether the buildings conform a perfect rectangle or not. Photographs by the author, drawings by Ambar Hernández

The two longest façades of the Governor's Palace (figs. 4a-b) run perfectly parallel to one another, while the two shortest ones deviate 2° from one another. The Nunnery Quadrangle (figs. 4c-d) does not form a perfect rectangle since its shortest sides deviate by 10°. As I expected, the orientation of a diagonal line across the courtyard is very nearly aligned with that of the solstice (70 az-250 az), and can be made equal to that of the base of the pyramid at Chichén-Itzá by slightly moving the farthest point of reference. This line is highly significant, as we will see later. The great pyramid of Uxmal also has nine platforms, still largely covered by the forest (evoking Catherwood's drawings), and despite the fact that it has a orientation similar to that of Kukulcán, the effect is not perceptible on its northern stairway due to its uneven contour.

Fig. 5. a, left) The Five-Storied Building viewed from the plaza. In front of it appears the platform where the missing pole stood; b, center) A view of the rounded eastern platforms. c, right) A view of the fly stairs. Photographs by the author

I expected to see the famous pole (or gnomon) of Edzná, 61 km. southeast of the city of Campeche, standing on its platform base (fig. 5a)and facing the Five-Storied Building (5-SB), but it was inexplicably removed.[5] Most scholars agree that the pole was used to calibrate the Mayan Calendar Long Count, when the zenith passage of noonday would fall on its surmounted capital-like top, casting a complete shadow on its body. Looking from the Nohochná stairs, the sun rises at both sides of the 5-SB, left and right, during the summer and winter solstices, respectively. In turn, the stela placed at the rear of the temple's second room (atop the 5-SB) is illuminated during the equinox sunset, a phenomenon that is still perceptible even though the vault of the first room has collapsed. Edzná overcame the lack of natural deposits of water by inventing the so-called *aguadas* system (washes). Water was collected from everywhere, even from the eastern side of the 5-SB, whose platform surfaces were rounded for this purpose (fig. 5b). This was an early manifestation of the architectural gospel, "Form follows function." Likewise, the innovative ramped stairs connecting the fifty rooms of the 5-SB suggest that the Maya were transforming the massive pyramids into habitable buildings (fig. 5c).

The constructive geometry of the Kukulcán pyramid

The extant Maya codices[6] contain calendar accounts, astronomical observations, social precepts, and political events, but no references to geometry. Unfortunately, we will never know if any texts on geometry had once existed, like those collected in the *Chilam Balam* for medicine.[7] As is well known, Diego de Landa ignominiously burned many Mayan writings after the Franciscan friars landed in the Yucatán Peninsula [de Landa 1986], an event that took place in Maní on July 12, 1562.[8] In spite of this shameful event, there are many unexplored places across the peninsula and, therefore, the birth certificate of our line could still be hidden somewhere. Meanwhile, only the buildings themselves can tell us what geometry lay behind their conception.

We have rounded off the measurements of the Kukulcán Pyramid because it is not perfectly symmetrical; all its sides and slopes deviate a little everywhere (fig. 6). The pyramid is comprised of nine platforms, each about 2.57 m. high (the actual height of the first one). Atop the last platform is a two-story temple 6 m. high, and thus the total height of the pyramid is 30 m. Each side of the pyramid's base is about 55.3 m. The average inclination of the platforms is 53°. The individual walls, or taluds,[9] incline by 72°-74°, and the slope of the stairways is 45°. Obviously, we cannot translate our metric measurements and degrees to the Mayan system because this is unknown. Yet we can wonder if some measurements were encoded in the pyramid, since numbers are everywhere. For instance, the number of steps of the pyramid totalled 365, corresponding to the number of days in a year,[10] while in turn, the 52 ornamental stone boards (all around the pyramid) coincide with the Mayan Calendar Round.[11] An architectural stone calendar seems to have ruled the program for the current pyramid.

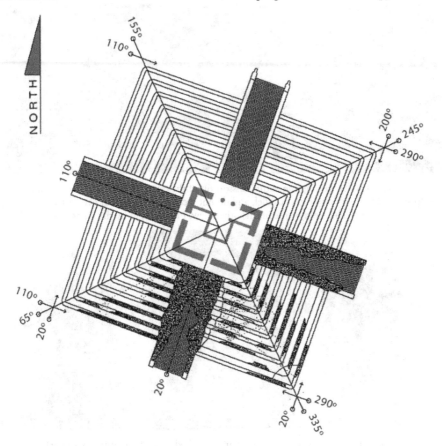

Fig. 6. Plan of the pyramid of Kukulcán in Chichen-Itzá. Drawing by Ambar Hernández based on the author's measurements

The sidewalks around the pyramid suggest that they were used to supply and handle all the materials needed to build the platforms progressively. During this process, low scaffolds were required by the stonemasons to give the platforms their typical talud form. In turn, as the size of the step's tread and riser is equal, the resultant slope of the stairways is 45°, which is unlikely to be calculated by an angular scale. If such a scale were used, why then do the platforms not show the same perfection of slope as the stairways? Standing atop the pyramid and looking downwards, it is evident that its four rounded corners do not align correctly, and that none the borderlines of the sidewalks line up tangentially either. This last phenomenon is a notorious mismatch since the borderlines do not recede (from the top to the base) accordingly in perspective (figs. 7a-b). Looking at the temple from the great plaza, it is quite intriguing to observe how its outer walls appear vertical when they actually slope outwards as they rise. Could this be an intentional visual refinement? It might be, or at least it suggests the use of perspective as a surveying method for building, in a way similar to what the Greeks did in the Parthenon.

It is widely believed that visual refinements were made in ancient cultures to keep the form of a building consistent.

Fig. 7. a, left) Disregarding the natural distortion of the camera lens, it is evident how the platforms' borderlines do not run parallel, and do not recede accordingly in perspective as well; b, right) The same is true in this other view. Photographs by the author

Fig. 8. a, left) The stairway of the older pyramid vaulted by a Mayan vault; b, center) At the entrance of the upper temple, a Chac-mool seems to be wondering, "Who dares to come here", as it protects the jaguar throne (at the rear) from unwanted visitors; c, right) A view of the stairway's right-hand balustrade without a serpent head at its foot. Photographs by the author

As we have mentioned, an older pyramid concealed beneath the current one, whose entrance is located to the northwest side of its stairway, was discovered during the 1930s. Literally, we have to go inside the present-day pyramid to find the main stairway of the older one. The "interior" stairway is roofed by a Mayan vault that leads to an upper sacred chamber, which remains intact just as it was originally found (figs. 8a-b). There, the silence of the past is a breathtaking experience. Even though the older pyramid has no serpent heads at its foot, the orientation of its base turns out to be almost the same as that of the current pyramid. The minimal deviation between the older stairway (110-112 az) and the current one (110 az) would not have nullified the effect, which leads me to hypothesize that the earlier pyramid exhibited the same effect of light and shadow (fig. 8c). In any case, a line of the new pyramid's base (what we call a diagonal) is oriented to the summer-winter solstice (summer solstice sunrise – winter solstice sunset, 65 az-245 az), as I confirmed by measurements on the site.[12]

The light and shadow effect

Were the Mayan builders capable of solving in 3D the projection between two non-coplanar planes? Was the light and shadow effect learned from the older pyramid, or was it noticed and reinforced during the construction of the current one? To solve a projection between two non-coplanar planes in 3D, one needs to have advanced knowledge of geometry. There is no evidence of such knowledge among the Mayans, not even that they were familiar with drawing instruments. This is why I am sceptical of all the scholars who believe that the effect was intentional, just as I am of the theory which suggests that the Quetzal chirps were engineered. This leads us to speculate if by chance the builders paid attention to the shadows cast when they were building the first or second platform of the new pyramid. If this actually happened, then no advanced geometry was necessary, but rather they worked directly with the real thing. This leads us to ask whether the platforms' deviation was a constructive defect or if the platforms were adjusted to line up with the triangular shadows.

Fig. 9. A 3D computer model of the pyramid showing four stages of the light-and-shadow effect during the equinoctial season. Drawings by Ambar Hernández

What we are certain is that the inevitable effect is produced by the interplay of two non-coplanar planes, namely: the northwest balustrade and the northwest dihedral angle of the pyramid (fig. 9). The inclination of each of these planes (45° and 43.5°, respectively) seems to have been determined independently. The fact is that the angular difference of 1.5° between the planes is what produces the illusion of the seven triangles moving downwards, an illusion that reaches its climax when the sun's rays hit the serpent's head at the foot of the stairway. If such a dazzling moment was indeed devoted to Kukulcán, then March 21 should be his celebration day. Kukulcán (also identified with Venus as coming down from heaven) was a god celebrated across the Yucatán Peninsula until the fall of Mayapán, a tradition thereafter held in Maní. Here, the people made a celebration lasting five days – from 16-Xul through 1-Yaxkin –, waiting for the last day when Kukulcán would come from heaven to receive their offerings [De Landa 1986: 98-99]. Such days, as deduced from Landa's correlation, would correspond to the early days of November, not March when the myth takes place in Chichén-Itzá. Aldana [2003: 33-51] correlates the celebration day with the first appearance of Venus in the evening sky of October 25, 1552, pretty close to Landa's correlation within a few days of difference.

Fig. 10. The southwest stairway of the pyramid of Kukulcán in precarious conditions.
Photograph by the author

The effect should also be visible during the equinox sunrise on the pyramid's southwest stairway, but its deteriorated conditions do not allow for this (fig. 10). Theoretically, it should be a perfect symmetrical effect: the morning ascent of the Plumed Serpent throughout the nine layers (represented by the 9 platforms) of the underworld (Xibalba), and its descent on the evening of the same day. However, neither the *Chilam Balam* nor the *Popul Vuh* can help to support this theory. Besides, why does not the southwest stairway have serpent heads, or should we be asking why those of the northeast stairway look mutilated?

The serpent heads at the foot of the northeast stairway were cut atypically, and look mutilated and superimposed (fig. 11a) instead of functioning as a true cornerstone like those of the Osario Pyramid, where the serpent heads are the original ones, and its balustrades lavishly ornamented (fig. 11b). It can be argued that both pyramids belong to different periods, despite the fact that they are about 350 meters away from each other, but it also can be argued that a constructive deflect cannot be justified by the use of a particular style. Landa reported having seen the serpent heads in place,[13] presumably those that Catherwood faithfully drew 300 years later, when they appear as cornerstones (fig. 11c).[14] Whether the original serpent heads were removed and then replaced with others (after the Stephens-Catherwood expedition) or whether they were removed and erroneously reinstalled during the stairway's restoration are hypotheses we should be examining.

Fig. 11. a, left) Architects and laymen alike can notice the wrong way in which the serpent heads were placed in Kukulcán's stairway; b, right) In contrast, the dexterous stonework of the Osario Pyramid's serpent heads, well integrated to the balustrades, makes evident the erroneous placement of those of Kukulcán. Photographs by the author

Fig. 11c. Catherwood's engraving of the serpents' heads

To conclude this section, the use of a rudimentary model to copy the older pyramid's effect, or engineering a new one, could be another theory, although remote. Disregarding the exact date on which the god Kukulcán was honored, such a model could have helped to visualize the pyramid and eventually learn from it the light and shadow effect. If such a model ever existed, it probably is buried somewhere in the pyramid.

A hypothesis about orienting the pyramid

The Ground-Penetrating Radar study conducted by Desmond [1996:23-30] has detected beneath the plaza a subsurface cultural world of which little is known. An extension of 250 m. by 400 m. of bedrock was filled up over a long period, ultimately buring a sacbé, several caverns, and previously laid down floors, all predating the construction of the plaza, an impressive undertaking for the time. The great plaza of Chichén-Itzá has an imperceptible slope, just enough to evacuate rainwater, which suggests that some surveying techniques were used for levelling and elevating it. Three main buildings occupy the great plaza: the Ball Court at the northwest, the Warrior's Temple at the northeast, and the Kukulcán Pyramid in the center. An Olmec hematite (a kind of lodestone compass), found in San Lorenzo (Veracruz), suggests the hypothesis of its use in planning Mayan centers, as Klokconík et al. pointed out [2007: 515-533].

However, the summer-winter solstice orientation given to the pyramid could have been intended to alert the people regarding the arrival of the season for sowing and harvest, and a lodestone compass would be useless for this purpose. Instead, I believe that a pole was used, since many buildings across the Mayan area seems to have a line oriented to the summer-winter solstice. Besides, it is easy to determine both the summer solstice sunrise – winter solstice sunset and winter solstice sunrise – summer solstice sunset orientations by observing the projecting shadow of a pole planted in the ground over the course an entire year – laying the equinox's orientation at the middle of the solstices' crossing shadows (see fig. 12).

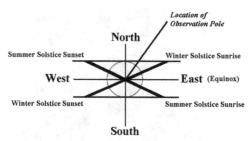

Fig. 12. Using the shadow cast by a pole to find the equinox orientations

Moreover, the pole was among the instruments for surveying that the Mayans presumably were familiar with; similar instruments are illustrated in the *Bodley Codice*.[15] A pole of wood rather than of stone (like that of Edzná, or that in front of the Governor's Palace in Uxmal) would be easier to construct, but less durable over time.

Let us suppose that the current pyramid's base was verified according to the summer solstice sunrise – winter solstice sunset orientation. In that case, three procedures could have been used to construct it: a) planting a pole on the temple's roof; b) placing a cross-staff (a surveyor's instrument for measuring offsets) at each corner of the older pyramid's summit; or c) using the older pyramid as a reference to lay out the new one. Let us examine each of these possibilities. A pole atop the roof of the older temple would have been useless because its shadow would not reach the ground at sunset. On the other hand, the cross-staffs could have helped surveyors prolong the base's diagonals at more distant points.

The third procedure seems the most logical one, based on the geometrical coincidences I found when superimposing the plans of both pyramids (the present-day one and the substructure). Thus, I decided to join the midpoints of the stairways' base of the present-day pyramid and see what happened; to my surprise, I found that the corners of the older pyramid's base perfectly circumscribe an auxiliary square (fig. 13).

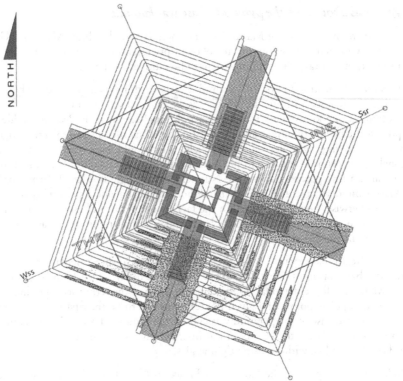

Fig. 13. The plan of the actual pyramid and that of the older one superimposed. Notice how an (imaginary) auxiliary square circumscribes the plan of the older pyramid. "The Line" is oriented according to the winter solstice sunset (wss) and summer solstice sunrise (ssr) orientation. Drawing by Ambar Hernández based on the author's measurements

What these matching plans are telling us is how the first auxiliary outline for the new pyramid's base was executed. Nothing else but simple tools such as henequen ropes (rope made from the fibre of the agave plant), several cross-staffs, and dexterous surveyors, were required to lay out the auxiliary square; but how? Let us imagine a team of surveyors standing atop the upper platform of the older pyramid, placing a cross-staff at each midpoint of the stairways' borderline, while on the ground, another team would do exactly the same at each base of the stairways. Thus, by lining up the upper cross-staff with the one on the ground, so as to have two visual points of reference, the surveyors would prolong lines on the ground until they intersected the auxiliary square that circumscribes the older pyramid's base.

In turn, the vertices of the auxiliary square turn out to be the midpoints for the new stairways. A trial and error process would lead both teams to equalize the sizes of the auxiliary square by manipulating the henequen ropes. In the same manner, the diagonals of the older pyramid could have helped to complete the layout of the new pyramid base. But how did the builders estimate the size of the temple's platform and the stairways' length? Re-examining fig. 13, it seems logical that an axis of the stairways was used for this purpose. Thus, tracing either the NE-SW or SE-NW axis on the ground, they could have divided it into three parts to set the dimensions of the stairways and the temple platform. In other words, the builders worked out the form of the pyramid on the ground.

A hypothesis about how the pyramid's base was laid out

The urban layout of some Mayan centers, such as Dzibilchaltun, Mayapán, Uxmal, Edzná, and Chichén-Itzá, seems not to follow a pre-established pattern, for all of them are different; in contrast, the buildings seem to be rationally oriented.

To start the construction of the pyramid, two basic things were needed: a chosen site, and a line of reference, our line. In practice, as we know, a square can be drawn from any line as long as one knows how to construct right angles. But if the Mayan builders did not possess the notion of a right angle, how did they construct a square from a single line without employing right angles? Most likely, they could have used what we call a diagonal as the prime line to set in place their buildings' foundation. It is not surprising to find such a line precisely across the Nunnery Quadrangle in Uxmal. In my opinion, the Mayan builders invented their own method for laying out a perfect square on the ground. Otherwise, how can we explain the fact that Kukulcán's base is almost a perfect square? (Three corners of the pyramid's base render perfect right angles, as I confirmed with measurements on site.)

As we already have pointed out, the summer-winter solstice orientation of a line of Kukulcán's base was most likely used as the prime line to lay out its foundation. In other words, what we called a diagonal (in the broadest sense of the term used to indicate a line joining two opposite points of a given figure) turns out to be the first laying out line by which the square base of the pyramid was laid on the ground. To sustain this conjecture, which in its own right is the history of our line, we have to prove that a square form can be laid out from a line without using right angles.

When I was inspecting the Governor's Palace in Uxmal, an idea came to me about constructing parallel lines without using right angles. Let us suppose that an oriented line L1 is laid on the ground where points a and b are marked by a rod planted at each one of them. Then, arcs of radius r1 are swinging by a rope attached to the foot of each rod, and where a line passes tangentially to the arcs, a parallel L2 to L1 is constructed.

Fig. 14. Experiment to prove the feasibility of what I call the Mayan method for laying out a square. We laid out the base of the older pyramid, on campus, without using right angles.
Photographs by Ambar Hernández

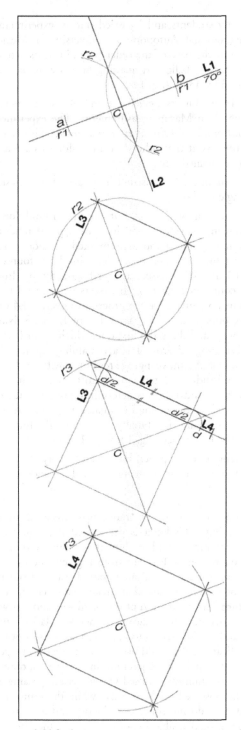

Fig. 14. The Mayan method for laying out a square. Geometric procedure by the author

In order to prove it, my students and I carried out an experiment in front of the Rectory building at the National Autonomous University of Mexico. We chose a spacious pavement to lay out the base of the older pyramid in its actual size. After four hours of work, using only ropes, chalk, and masking tape, we succeeded in constructing an almost perfect square of 33 x 33 m. (fig. 14).

Of course, we assumed the 33 m. as a given datum, since we cannot know how it was determined or what it means in Mayan measurements. The experiment was based on what I have called the *Mayan method for laying out a square*. This is a method that can be useful for explaining the layout that other Mayan buildings have as well. Its general statement and solution are explained next.

Problem: Given an oriented line L1, construct a square of a pre-established side L4 without using the right-angle method.

Solution: Let line L1 be a straight line set out on the ground (oriented by a pole according to the summer-winter solstice), and let c be a point randomly set along it. Carrying out radius r1 from c, points a and b are marked out along L1. Then, swinging arcs from a and b, with a radius r2 > ab/2, a straight line L2 is found where these arcs intersect each other. As L1 and L2 intersect at c orthogonally, by dropping a circle of radius r2, a L3 square is conformed where r2 intersects with lines L1 and L2. Now, to construct from the random L3 a given L4, we proceed to overlap L4 onto L3. In doing this, we would have three options: L4 > L3, L4 < L3, or L4 = L3. Naturally, the latter would occur once in a thousand. Taking the first one, which is our figure's case, carrying out L4 onto L3 and subtracting L3 from L4, a remaining segment (d) would result, which in turn is divided in half. Again, swinging arcs but now from the extreme points of L3, of a radius r3 = d/2, and sliding L4 tangentially to these arcs, it would fit exactly at the intersection points with L1 and L2, thus determining one L4 side of the L4 square. To trace the remaining three L4 sides of the L4 square, it would suffice to carry out the new radius r3 until it crosses L1 and L2. Finally, by joining the found points, a perfect square of the given side L4 is determined (fig. 14: steps 1 through 4).

Due to the restricted length of this paper, both the corollary (L4 < L3) and the proof of the theorem behind this problem will be treated formally in another paper.

Conclusions

Perspective as representation, in my opinion, was preceded by perspective as a surveying method for building. This latter is the underlying issue we have speculated about in this article, in the attempt to find out how it was employed in the Mayan world. We have proved that a square can be laid out from a given line by a simple surveying method. Most likely the base of the Kukulcán Pyramid was laid out in that way. It is precisely the application of such a method that makes the orientation of the pyramid's base so unique. And, in turn, the inclination of the platforms and stairways, ruled by the constructive process of the pyramid, introduced another singularity. On the other hand, the special effect of light and shadow seems to take place as a result of these singularities; there is no evidence to indicate that it could have been engineered by geometrical means at the time. Furthermore, by modelling the pyramid in 3D, we can observe that the effect is still visible even when its orientation is varied within certain parameters. This explains why seven triangles can be observed on March 21, while the number of triangles seen varies between three and nine during some days before and after the equinox. Lastly, thinking as a builder, the most plausible hypothesis is that the effect was noticed and reinforced during the pyramid's construction (as opposed to having been pre-planned). At least, that is the one I am in favor of, until it can be proved otherwise.

Notes

1. Here, the word 'discovered' most likely indicates 'a place to settle was found'.
2. See http://en.wikipedia.org/wiki/Chichen_Itza.
3. Eric S. Thomson boasted that his knowledge of Maya was self-taught. At least he has only himself to blame for what he wrote: "I was given charge of the excavations of the Caracol (the Observatory), that queer ugly round building which…" See *Maya Archaeologist* (1963, University of Oklahoma Press, p. 37).
4. Azimuth (az): In plane surveying, a horizontal angle measured clockwise from north meridian to the direction of an object or fixed point. All azimuth measurements quoted in this paper are the author's.
5. In December 2007, the pole was not there. This pole is a tapered shaft of stone surmounted by a capital-like top. According to the National Institute of Anthropology (INAH) of Mexico, it was mistakenly aligned at the northeast façade of the Five-Storied Building during the restoration of the site, and for this reason, it was put away to avoid erroneous interpretations. This was communicated by INAH authorities in an official letter addressed to the author (Feb. 26, 2008/ Of. Núm. 401-7/333).
6. The only extant Mayan Codices (folding-screen books) are: the Dresden, Paris, Madrid, and Grolier.
7. There are at least three books on medicine: The *Chilam Balam* of Káua, the *Chilam Balam* of Tekax, and the *Chilam Balam* of Nah.
8. *Usaba también esta gente de ciertos caracteres o letras con las cuales escribían en sus libros sus cosas antiguas o ciencias, y con estas figuras y algunas señales de las mismas, entendían sus cosas y las daban a entender y enseñaban. Hallámosles gran número de libros de estas sus letras, y porque no tenían cosa en que no hubiese superstición y falsedades del demonio, se los quemamos todos, lo cual sintieron a maravilla y les dio mucha pena* [De Landa 1986: 104-105].
9. A talud is an outer wall that slopes inward as it rises. The stonework in flint and obsidian is remarkable.
10. By adding the steps of the four stairways plus the step of the temple's base, the result would be $[(4 \times 91) + (1)] = 365$.
11. It was a Mayan custom to renew almost everything every 52 years, which explains the existence of many superimposed structures everywhere in their cities.
12. The azimuthal alignments presented here were taken in situ by the author with a Brunton Transit (surveying compass adjustable for magnetic declination; azimuth accuracy of $\pm \frac{1}{2}°$ with 1° graduations). To sustain the theory of the solstice line being used by the Mayan to either build squares (Kukulcán) or rectangles (Uxmal), the author needed to rely on his own architectural data of the places he visited. Despite the fact that the author's alignments might differ from other alignments already published, they are relatively congruent among themselves, and therefore they are to be considered accurate. Chichén Itzá location: latitude 20° 40' 56''' N, longitude 88° 34' 05'' W. As it is known, the magnetic declination (MD) varies with the passage of time all over the world. The author estimated a MD of 1° 30' E, based in the 2007 isogonic chart for North America, while for December of the same year, the National Geophysical Data Center (NGDC) estimated a MD of 0° 30' E.
13. *Había, cuando yo le vi, al pie de cada pasamano, una fiera boca de sierpe de una pieza bien curiosamente labrada. Acabadas de esta manera las escaleras…* [De Landa 1986: 113].
14. Catherwood's engraving depicts a neat stone cut, perpendicular to the neck of the serpent heads, not in diagonal position as present-day heads have; cfr. [2000: 357, fig. 14].
15. The *Codex Bodley* [c. 1500] is a pictographic manuscript of the Mixtec culture (south-west from the Maya area). Among the illuminated glyphs it contains (carefully organized on a deer skin of 22 feet long by 10 inches wide), there is a "X" conformed by crossed sticks, and a "V" with an eye in the middle, suggesting the presence of surveying instruments to build. See [Codex Bodley: 15, 19, 21, 32].

References

ALDANA, Gerardo. 2003. K'uk'ulkan at Mayapán: Venus and postclassic Maya statecraft. *Journal for the History of Astronomy* 34 (114): 33-51.

BARRERA Vázquez, Alfredo and Silvia RENDÓN, trans. 2005. *El libro de los Libros de Chilam Balam*. Mexico, Fondo de Cultura Económica.

CODEX BODLEY. c. 1500. Bodleian Library, University of Oxford, Codex Bodley 2858. http://www.famsi.org/spanish/research/pohl/jpcodices/bodley/index.html

DE LANDA, Diego. 1986. *Relación de las Cosas de Yucatán*. Mexico, Ed. Porrúa.

DECLERCQ, Nico F., Joris DEGRIECK, R. BRIERS, and O. LEROY. 2004. A theoretical study of special acoustic effects caused by the staircase of the El Castillo pyramid at the Maya ruins of Chichen-Itza in Mexico, in: *The Journal of the Acoustical Society of America* 116 (6):3328-3335.

DESMOND, Lawrence G. and William A. SAUCK. 1996. Entering the Maya underworld: A ground penetrating radar survey at Chichen Itza, Yucatan, Mexico. Pp. 23-30 in *Innovation et Technologie au Service de Patrimoine de l'Humanite*, Actes du colloque organise par Admitech en collaboration avec l'Unesco, Paris, June 24, 1996.

GARCÍA-SALGADO, Tomás. 2004. *Instrumentos para la Geometría Perspectiva* (*Instruments for Perspective Geometry*). Mexico, FA, UNAM, pp. 40-41.

KLOKOČNÍK, Jaroslav, Jan KOSTELECKÝ, and František VÍTEK. 2007. Pyramids and Ceremonial Centers in Mesoamerica: Were They Oriented Using a Magnetic Compass? *Prague, Studies in Geophysics and Geodesy* 51: 515-533.

LUBMAN, David. "The Mayan Pyramid". http://www.ocasa.org/MayanPyramid2.htm.

STEPHENS, John L. 2000. *Incidents of Travel in Yucatan*. Mexico, Ed. San Fernando.

TEDLOCK, Dennis, trans. 1996. *Popol Vuh: The Mayan Book of the Dawn of Life*. New York, Touchstone.

About the author

Tomás García-Salgado received his professional degree (1968), Master's degree and Ph.D. (1981-1987) in architecture. He is a formal researcher in the Faculty of Architecture of the UNAM (México), and holds the distinction as National Researcher, at level III. Since the late 1960s, he has devoted his time to research in perspective geometry, his main achievement being the theory of Modular Perspective. He also has several works of art, architecture, and urban design. More information regarding his work is available at http://perspectivegeometry.com.

John G. Hatch

Department of Visual Arts
The University of Western Ontario
London, Ontario
Canada N6A 5B7
jhatch@uwo.ca

Keywords: El Lissitzky, Theo Van
Doesburg, De Stijl, Relativity,
relationships between art and science

Research

Some Adaptations of Relativity in the 1920s and the Birth of Abstract Architecture

Abstract. John Hatch examines the friendship between Theo Van Doesburg and El Lissitzky, which was fuelled by a shared interest in scientific theories. Both moved from painting to architecture in seeking out a form best suited to conveying the spatiotemporal experiences phrased by Relativity, resulting in some remarkably innovative architectural designs and theories.

El Lissitzky set out to become an artist, but after failing the admittance test at the Saint Petersburg Academy of Arts, he turned his attention to architectural engineering, graduating from the Technical Institute of Darmstadt in Germany. Upon his return to Russia in 1914 Lissitzky followed up on his interest in art, designing and illustrating children's books, most notably of Jewish folktales. Lissitzky's talents in engineering and art resulted in his being hired in 1919 as head of the Workshops of Graphic and Printing Arts, and Architecture at the Artistic-Technical Institute in Vitebsk (Belarus), an art school established by Marc Chagall. It is there that Lissitzky met the Ukrainian painter Kazimir Malevich whose Suprematist works would have a profound impact on Lissitzky's career as an artist.

A notable aspect of Malevich's art, for Lissiztky, was its incorporation of scientific theory. Malevich drew on thermodynamics, describing the coloured forms of Suprematism as representing nodes or concentrations of energy, and its whole narrative as one paralleling the universe's evolution toward thermal death, as postulated by the second law of thermodymanics. The White on White series of 1917-1918, represents the penultimate moment of the end of the material world for Malevich, in favour of a higher spiritual reality inspired by his interest in theosophy [Hatch 1995: 120-168]. Lissitzky did not share Malevich's spiritualist beliefs and where Malevich saw Suprematism as a terminal point for human history, Lissitzky saw it as the starting point for a complete transformation of our material existence. Lissitzky wanted to use Suprematism, or his variant of it, the "Proun" (acronym of "project for the affirmation of the new"), as a blueprint for social reconstruction – a hope fuelled by the October revolution of 1917.

In devising his own variant of Suprematism, Lissitzky wanted to update the science it drew upon. One of the earliest and most obvious examples of the incorporation of new scientific theories in Lissitzky's work is found in *Proun G7* (fig. 1). This work is based largely on a diagram found in Hermann Minkowski's seminal essay "Space and Time," published in 1908 (fig. 2).[1] This is even more clearly illustrated in the studies for *Proun G7*, where the copying of Minkowski's space/time continuum diagram is quite literal. The most telling aspect is that *Proun G7* not only incorporates the hyperbolas found in Minkowski's diagram, it transcribes the oblique presentation of the *x*- and *y*-axes as well.

Fig. 1. El Lissitzky, *Proun G7*, 1923, tempera, varnish and graphite on canvas, 77 x 62 cm, Kunstsammlung Nordhein-Westfalen, Düsseldorf

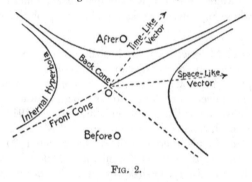

Fig. 2. Hermann Minkowski's space/time continuum diagram, Fig. 2 from the essay "Space and Time" [Minkowski 1908: 84]

Minkowski was a Russian-born German mathematician who was a teacher of Einstein's in Switzerland. He was one of the earliest scientists to appreciate the full potential of Einstein's theory, and his text supplied the first rigorous mathematical/geometric treatment of Relativity. Minkowski endorsed Einstein's concept that our perception of reality is invariably associated with the four dimensions of space and time, remarking that "Nobody has ever noticed a place except at a time, or a time except at a place" [Minkowski 1908: 76]. His particular formulation of this idea involved the use of non-Euclidean geometry and imaginary numbers. Minkowski's analysis concluded by calling on science to finally abandon the classical notion of absolute space in favour of relative spaces, or as Minkowski himself put it: "We should … have in the world no longer space, but an infinite number of spaces" [Minkowski 1908: 79].

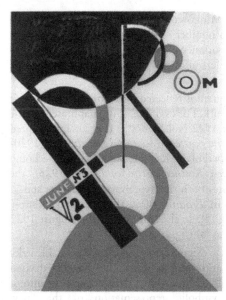

Fig. 3. El Lissitzky, Cover of the art magazine
Broom, vol. 2, no. 3, June, 1922

Fig. 4. El Lissitzky, *First Kestner Portfolio,
Proun*: print no. 3, 1923, lithograph, 64.0 x
49.0 cm

Fig. 5 (above). El Lissitzky, *Proun 43*, c.
1922, watercolour, gouache, india ink,
aluminium paint, collage, on board, 66.8 x
49.0 cm, State Tretiakov Gallery, Moscow

Fig. 6 (right). El Lissitzky, *MA: Cover Proof*,
1922, linocut on transparent paper, 27 x 19.8
cm, Municipal Van Abbemuseum, Eindhoven

This is a passage which must have endeared Minkowski to Lissitzky, since the latter's Prouns represent an attempt at working with a number of different spaces. Surprisingly, despite extensive scholarly mentions of the importance of Einstein's Theory of Relativity for Lissitzky, no one has ever suggested the derivation of *Proun G7*, and subsequent works, from Minkowski's diagram.

The overall importance of Minkowski's diagram is reflected in the number of times it reappears, under various guises, in Lissitzky's work. For example, it is paraphrased on the cover of the art magazine *Broom* from June of 1922 (vol. 2, no. 3) (fig. 3), and adapted for the image of "the new," found in the *Figurine portfolio, Victory Over the Sun* (1920-21). In both cases, we find Lissitzky using hyperbolas which are slightly offset, echoing Minkowski's oblique presentation. In the *First Kestner portfolio, Proun* (1923), the image found on sheet number three incorporates both the hyperbolas and the *x*- and *y*-axes found in Minkowski's diagram (fig. 4). In *Proun 43* (ca. 1922) Lissitzky simply transposes the image found in *Proun G7* (fig. 5). It also appears on the August 1922 cover of the Hungarian magazine *MA* (vol. 7, no. 8) (fig. 6).

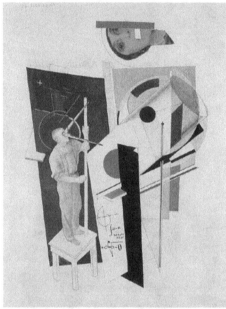

Fig. 7. El Lissitzky, *Tatlin, Working on the Monument*, 1921-22, collage, (29.2 x 22.9 cm), Grosvenor Gallery, London

Significantly, in all of these, Lissitzky anthropomorphizes the diagram devised by Minkowski, transforming it into a symbolic representation of the "new man." In both *Proun G7* and *Proun 43* there are a number of elements near the focal point of each image which are referred to by the art historian Alan Birnholz as "architect's equipment" [Birnholz 1973: 150-152]. These likely refer to Lissitzky's own architectural training. In turn, they also point to the fact that Lissitzky's "new man" was the architect or constructor of a new reality, one founded on the new mathematics and the Theory of Relativity. This point is more clearly made in Lissitzky's collage *Vladimir Tatlin, Working on the Monument* (1921-22) (fig. 7), in which Lissitzky includes a mathematical formula composed of an imaginary number (the cubic root of -0) and a symbolic expression of positive/negative infinity.

It is most likely a symbolic reference to Minkowski, who not only used imaginary numbers in his equations dealing with the space/time continuum, but also defined the continuum as extending from negative infinity to positive infinity. Thus, the inclusion of this mathematical formula in his collage suggests that Lissitzky saw the Russian sculptor Tatlin as an embodiment of the "new man," although there is little evidence of Tatlin's interest in Relativity.

While working on the Tatlin collage at the end of 1921, Lissitzky was in Germany where he met the Dutch artist Theo Van Doesburg, who at the time was trying to obtain a teaching position at the Bauhaus in Weimar. Van Doesburg was the driving force of De

Stijl, the Dutch modern art movement that was founded in 1917 by Van Doesburg, Bart van der Leck, and Piet Mondrian. Curiously enough, Lissitzky and Van Doesburg shared a common interest in science, and where Lissitzky's art was influenced by Malevich, Van Doesburg's mature work was fuelled by his passionate interest in Mondrian's painting. The similarities continue in an eerie fashion. Mondrian was also deeply involved in spiritualist beliefs and shared Malevich's interest in theosophy. Like Malevich, Mondrian incorporates a number of elements from nineteenth-century science, in part due to theosophy's adaptation of nineteenth-century scientific theories, although not as thoroughly as Malevich. Mondrian makes only the occasional but nevertheless significant references to energy and matter, and specifically the concept of the ether. Like Lissitzky, Van Doesburg did not share as passionate an interest in the spiritualism of his mentor, and wanted to update the scientific references.

Van Doesburg's interest in the physical sciences emerges around 1918 and is related to one issue on which he and Mondrian could not see eye-to-eye, namely the representation of time or movement in painting. Again, like Malevich, Mondrian's work was a signpost to a higher spiritual dimension, one which was immutable and absolute, and to include time in the realm of the timeless was obviously not an option for Mondrian. Like Lissitzky, Van Doesburg was more interested in how to translate Mondrian's visual idiom into practical, material terms, and consequently, time/movement was integral.

Van Doesburg wholeheartedly embraced the premise that all is relative, all is in continual movement and, consequently, that there are no absolutes in the universe. He had discussed these ideas with Mondrian just before the latter left for Paris in 1919. In a letter to the De Stijl architect J.J.P. Oud, Van Doesburg related that he had met with Mondrian in June of 1919 and discussed at length his belief that all is in '*mouvement perpetuel*.' He added that Mondrian rebutted his interpretations in a rather dogmatic manner. Shortly thereafter, no mention is found in Van Doesburg's writings for *De Stijl* of these concepts, except under the guises of I.K. Bonset and Aldo Camini. These two authors were pseudonyms for Van Doesburg and represented, in part, an outlet for his ideas on time and space. This appears to have been done specifically to avoid offending Mondrian, as Carel Blotkamp suggests [1986: 30]. Not surprisingly, Mondrian warned Van Doesburg about the inclusion of these contributors in *De Stijl* [Blotkamp 1986: 30].

Bonset claimed to be a Dada poet and, as such, his inclusion in De Stijl seemed rather odd. Yet reading his work, we discover that he explicitly voiced Van Doesburg's views on space and time. In his "X-Images" (published in *De Stijl* issues of May and July 1920) a debt to scientific thought is immediately apparent in the title of these poems, where *x* suggests not only a link to x-ray imagery, but also to an unknown quantity in mathematical equations: an interpretation which makes sense in terms of the emphasis Van Doesburg would place on mathematics and geometry in his painted works. Furthermore, space and time are also themes, as expressed in the following passage:

> did you experience it ph y S I C A L L Y
> On
>   -space and
>   -time
> pastpresentfuture
> the behindhereandyonder
> the mix-up of the nought and the phenomenon.[2]

But why Dada? What attracted Van Doesburg to Dada was, firstly, its destructive character as an eradicator of past tradition. This was a goal Mondrian endorsed as well and is why in letters between the two, around 1920, they would sign themselves as "Dada-Does" and "Dada-Piet" [Holtzman & James 1986: 124]. But for Van Doesburg, Dada was more than simply a needed cultural bulldozer, it also represented a new image of reality which he himself adopted; it was an image founded on relativistic principles. As Bonset related in "What is dada???????", published in *De Stijl* in 1923, "Dada is the great phenomenon which is parallel to the relativistic philosophies of the present period … .Dada cannot be fixed by laws" [Bonset 1923: 131]. Consequently, Bonset claimed Einstein as a dadaist.

This position that reality cannot be fixed by laws later became a fundamental principle of Van Doesburg's "Elementarism", the term he used to describe his new art, which incorporated the oblique in opposition to Mondrian's orthogonal. As Van Doesburg related in 1927, "Elementarism advocates the complete destruction of traditional absolutism"; he added that it "acknowledges a form of plastic expression in four dimensions, the realm of space-time" [Van Doesburg 1927: 163, 165].

What is interesting about these excursions by Van Doesburg into the realm of relativistic philosophy is that they belie their varying sources. It would be a mistake to claim that Van Doesburg knew much of Einstein's Theory of Relativity before 1921. In fact, the first explicit reference made to Einstein only occurs in 1923 in the article "What is dada???????". Einstein was not the only individual to formulate a relativistic theory at this time. Too often we jump to the conclusion that if time and space are mentioned together, a reference is being made to Einstein's theory. This is certainly not the case with Van Doesburg. The references Van Doesburg made to space and time in his writings between 1913-1918 are taken from Theosophic texts. For example, Van Doesburg was particularly enthusiastic about M.H.J. Schoenmaekers's works, a number of which he had read by 1918. In his book *The New Image of the World* (1915), Schoenmaeker sets forth a space/time theory based on an interpretation of fourth-dimensional theories current at that time [Blotkamp 1986: 30]. It is only as of 1918 that Van Doesburg began to examine scientific texts dealing with relativity.

Van Doesburg may have turned to scientific interpretations of relativity in 1918, but nothing suggests that he specifically studied Einstein's Theory of Relativity. In a letter dated Sept. 22, 1918 to the poet Antony Kok, Van Doesburg wrote that he had read Henri Poincaré's *New Mechanics* and E. Cohn's *The Physics of Time and Space* and, furthermore, recommended that Kok read "the Relativity theory of Professor Lorentz." The latter is a rather curious statement which has never been questioned before. It has probably been assumed that Van Doesburg meant Lorentz's texts dealing with Einstein's Relativity theory. But he may have simply been referring to Lorentz's own Relativity principle. It is not commonly known outside the discipline of the history of science that Lorentz had, with the assistance of Poincaré, formulated a principle of relativity. In general terms, the differences between the Lorentz/Poincaré principle and Einstein's theory are not obvious, but in specific terms Einstein's theory was more extensive in its bringing together mechanical and electromagnetic phenomena, whereas Lorentz and Poincaré's formulation stressed the mechanical to the exclusion of the electromagnetic. Also, Einstein found no use for the ether, replacing it with space, while Lorentz and Poincaré retained the concept of an ether.

The reason why Van Doesburg's first encounter with relativity would have been with the Lorentz/Poincaré principle rather than Einstein's theory could simply be because Lorentz was Dutch. Also, Bart van der Leck, the co-founder of De Stijl, is known to have attended lectures given by Lorentz and he may have related some of Lorentz's ideas to Van Doesburg [James 1957: 60]. Another factor may have been Henri Poincaré's popularity with artists at this time. He's mentioned by the Italian Futurists, New York Dada, and the Cubists; Van Doesburg published Poincaré's article "Pourquoi l'espace à trois dimensions" in *De Stijl* magazine and lists Poincaré's *New Mechanics* in the library of *De Stijl* [Blotkamp 1986: 29-30]. This is not to say that Van Doesburg was not aware of Einstein's theory, but it is possible that Van Doesburg thought that the Einstein and Lorentz/Poincaré theories were essentially the same; a mistake made by a number of contemporary physicists as well, most notably Max Planck [Hirosige 1976: 70].

Van Doesburg's interest in coupling time and space would manifest itself in earnest in 1921 and, significantly, in architecture. And in a series of lectures given in Weimar in 1921, he announced his new vision for architecture:

> In contrast to the painterly approach inherent in an architecture of two-dimensional facades, the task of the architect is to annul three-dimensional volume by correctly expressing the relationships involved in the arrangement of space ...
> ... For modern architecture the proper use of colour in space is the most important and difficult issue of our time ... .A balance between the elements of space and time can be achieved only in terms of coloured plasticism, which is to say, in terms of painted three-dimensional space-compositions [Van Doesburg 1922: 124-125].

It is in his architectural designs that Van Doesburg would fully flush out his ideas on space and time in art, which would eventually find their way into his paintings.

Lissiztky did not turn to architecture as readily as Van Doesburg did in giving form to ideas related to relativity. However, he did abandon Minkowski's diagram as a Proun image by 1923. One suspects that Lissitzky may have been initially attracted to Minkowski's diagram because it was one of the only "images" of Relativity available. Unfortunately, as Lissitzky must have realized, it is a symbolic rather than actual depiction of the unity of time and space. If the Proun works were to be an architectural blueprint for a new reality, they had to propose how that reality could be realized as material form. This could obviously not be achieved with Minkowski's diagram of the space/time continuum. Consequently, Lissitzky continued to focus his attention on the manipulation of space, which had always been a central component of the Proun works, but with the added element of time. But this introduced a huge problem, namely of how to translate time or motion in a static medium like painting.

Motion is one of the elements involved in Lissitzky's *Proun 93: Free-Floating Spiral* (c. 1923) (fig. 8). The spiral is in fact a series of concentric circles placed one inside the other, with the center point of each moving progressively closer to the bottom left portion of the largest circle. Visually, there are two effects generated by this design. Firstly, the close proximity of the lines gives the illusion that the spiral is in movement: almost like the vibration of a metal spring. Secondly, the spiral creates the impression of a cone, but one which is ambiguous as to whether we are looking at the outside of the cone or looking into it. These effects were certainly directed toward creating an optical sensation of the unity of time and space, i.e., movement in space.

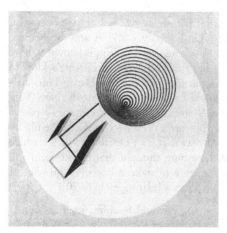

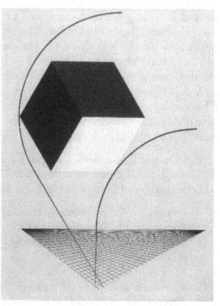

Fig. 8 (above). El Lissitzky, *Proun 93: Free-Floating Spiral,* c. 1923, graphite and colour pencil, india ink, pen and gouache, 49.9 x 49.7 cm, Staatliche Galerie Moritzburg Halle

Fig. 9 (right). El Lissitzky, *Proun,* 1924-1925, pen and ink, watercolor, collage 64.6 x 49.7 cm, Museum of Art, Rhode Island School of Design

The optical effect produced in *Proun 93* is recreated in *Proun* (1924-1925) (fig. 9) and *Proun 99* of 1925. The latter painting was Lissitzky's last and may have been conceived as such since it is the simplest and most succinct summary of the Proun theme. In *Proun 99*, Lissitzky creates a multi-dimensional experience involving one-dimensional lines, a two-dimensional strip, a three-dimensional cube and what is most likely a non-Euclidean grid. The latter is suggested by the slight curvature of the grid which, given Lissitzky's familiarity with Carl Friedrich Gauss, is most likely a reference to Gauss's co-ordinate system, which was designed to solve irregular grids such as those describing a curved surface. Although each of the dimensional components found in *Proun 99* appear to generate a coherent image, a closer examination shows otherwise. The one-dimensional lines appear to support the three-dimensional cube, yet by definition this is impossible. In terms of the two-dimensional strip, the viewer reads it at the top as being a certain distance from him/herself, but as the eye moves down this strip, its spatial position vis-à-vis the viewer changes when examined in relation to the cube, and changes yet again in relation to the grid. The spatial position of the cube is also ambiguous: the two lines appear to situate the cube at the edge of the grid, yet the size of the cube and the position of the grid itself suggest otherwise. This play on dimensions is typical of all Lissitzky's Prouns, and was directed at making us aware of the nature of space.[3]

Proun 99 illustrates what Lissitzky would call imaginary space, namely the unity of space/time through the three-dimensional cube which Lissitzky manipulates in a similar manner as the spiral found in *Proun 93.* The application of silver gray paint on one side of the cube creates a shimmering surface, paralleling the vibration of the spiral in *Proun 93.* The cube, like the spiral, is also handled in such a way as to allow a dual reading: either we are looking into the cube or the outside of it (an effect which is accomplished, in part, by the slight irregularity of the cube). Lissitzky hoped that this dual reading would create the impression of an inward/outward-shifting cube, and thus generate the illusion of spatial movement in painting. With this illusion and the dialogue between the different dimensions presented in *Proun 99*, Lissitzky recapitulated a key objective of *Proun* which he outlined in 1921, when he wrote:

> Proun advances towards the creation of a new space, and by dividing it into the elements of its first, second and third dimensions passing through time, it constructs a polyhedral but uniform image of nature [Lissitzky 1976: 70].

Proun 99 was a successful work, but one suspects it was not a satisfactory one for Lissitzky. The shifting cube was essentially an optical trick, an illusion, which represented a reversion to the trompe l'oeil devices of the Renaissance. It simulated movement rather than generating real movement. Lissitzky himself noted in his article "A. and Pangeometry" [1968] (a tribute to Nikolai Lobachevski's famous essay of 1855) that Futurism and Suprematism presented only static symbols of movement and that the new art had to finally cross the threshold of incorporating real motion, real time. This is the most likely reason why Lissitzky finally abandoned painting. As an inherently static medium, it was clear that painting was unsuited for the task Lissitzky ultimately had in mind. But his abandonment of painting was certainly not a tragic decision for Lissitzky, since he did present Proun as the "interchange station between painting and architecture." Van Doesburg may have had a hand in this move.

Van Doesburg's knowledge and use of physical theories grew immensely after his first encounter with Lissitzky at the end of 1921. Obviously, Van Doesburg could not have found a more fitting fellow enthusiast of modern science: someone well-versed in the finer points of physics and mathematics, and particularly Einstein's Theory of Relativity. Lissitzky appears to have convinced Van Doesburg that Einstein's Theory was worth considering more fully. Whether through extensive discussions with Lissitzky or a more serious examination of the literature, Einstein's Theory begins to play a more considerable role in Van Doesburg's work by 1922. This is revealed in the Dada passages quoted earlier, as well as the new approach to architecture Van Doesburg outlined while teaching in Weimar.

In 1922-23, Van Doesburg collaborated with a young Dutch architect, Cornelius van Eesteren, whom he met in Wiemar in 1922. Van Eesteren was an architecture student whose final student project was the design of a university for Amsterdam. He had sketched out the design in the Netherlands, but it underwent a radical transformation after he met Van Doesburg. The actual extent of Van Doesburg's contribution to the design of the structure is unknown and problematic. He did claim a substantial role, as he noted in a letter to Van Eesteren: "I could point out to you … your University before and after that first stay in Weimar". Although the final appearance of the structure is well-known, Van Eesteren's original conception is not. But given the traditional, classical design of Van Eesteren's earlier projects, the final design of the central building of Amsterdam University appears to have been influenced substantially by Van Doesburg. I am dwelling on this point since one of the striking aspects of the plan is its X shape (fig. 10), a feature which reappears in another collaborative effort, "La Cite de Circulation" (1924-29). This recalls Lissitzky's "New Man" and its derivation from Minkowski's diagram. It seems reasonable to suggest that it inspired the Amsterdam University plan, especially when one compares Lissitzky's image of the "New Man" published in *MA* and Van Doesburg's *Composition for the Floors* (1923) (fig. 11) and *Colour Designs for the Ceilings* (1923) (fig. 12).

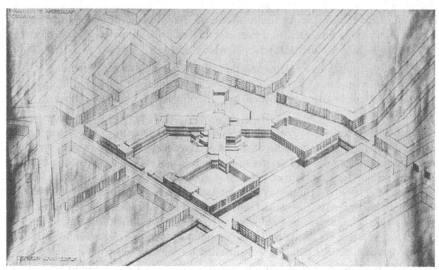

Fig. 10. Cornelis van Eesteren with Theo van Doesburg, *Design for a University in Amsterdam*, 1922, Van Eesteren-Fluck & Van Lohuizen Stichting Foundation, The Hague

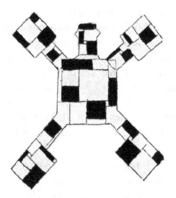

Fig. 11 (above). Theo van Doesburg, *Composition for the Floors*, 1923, fig. 12 in *L'Architecture Vivante*, no. 9, autumn 1925

Fig. 12 (right). Theo van Doesburg, *Colour Design for the Ceilings*, 1923, fig. 12 in *L'Architecture Vivante*, no. 9, autumn 1925

This may have been one way Van Doesburg translated Relativity into architecture. Significantly, it is Van Doesburg's designs for the ceiling of the University Hall that inspired his counter-compositions involving the use of the oblique as opposed to Mondrian's orthogonal relationship.

Lissitzky may also have related to Van Doesburg the interpretation of the past, present and future derived from the Theory of Relativity, a concrete example of which found its way into the work of Russian sculptor Naum Gabo. This interpretation of events in a Relativistic universe is outlined in Minkowski's essay "Space and Time" and

---

explained at length by Hermann Weyl,[4] whose "Light Cones" Gabo copied almost literally [Hatch 1995: 358-364]. It is quite likely that Lissitzky had read Weyl and was familiar with Weyl's "light cones", which distinguish between the active future and passive past. This interpretation found its way into Van Doesburg's 1924 article, "Surrealism. Realistic Dialogue" published in *De Stijl* (1924), in which, in his discussion of Cubism, he dissects himself as narrator into "my past I" and "my future I". It is also in this text that Van Doesburg noted his dissatisfaction with the intuitive approach to art, wishing to replace it with a scientific (mathematical) determination [Baljeu 1974: 68-70]. It may be far-fetched to suggest a parallel between this text and Relativity's interpretation of time, but given the close relationship between Lissitzky and Van Doesburg the suggestion seems justified.

In 1923, Van Doesburg was given the opportunity to express his architectural principles fully and concretely; previous to this, all of Van Doesburg's architectural experiments were applied to already constructed buildings. The art dealer Léonce Rosenberg, who sold Mondrian's works and was a supporter of De Stijl, commissioned the group to construct a villa for him. The buildings, three in all, were never constructed, since Rosenburg did not have the funds to build them. But the plans and models provided an important experimental ground that allowed Van Doesburg, with the help of Van Eesteren, to refine and further develop his new ideas on architecture.

Fig. 13. Theo van Doesburg, *Countercomposition in Primary Colours for an Artist's House*, 1923, 36.8 x 38.1 cm, Dienst Verspreide Rijkskollekties, The Hague

In the Rosenberg commission, Van Doesburg draws upon Mondrian's concept of planar construction (fig. 13). Mondrian had defined Neo-Plastic architecture as based on "a *multiplicity of planes*" [Mondrian 1922: 171]. This is not how Van Doesburg had conceived his earlier architectural models. One of the first major manifestoes dealing with the Rosenberg commission, published in *De Stijl* in 1924, related that the structures were formless, based on the definition of space by way of rectangular planes. These planes defined a system of spatial relationships, where no one element of the construction could be viewed as a closed, inactive space. This definition meant the rejection of the groundplan in architectural design, as Van Doesburg remarked:

> The ... planes, which separate the different functional spaces, can be *mobile*, which means that the separating planes ... can be replaced by movable screens or slabs ... .In the following phase of this development in architecture, the groundplan must disappear completely. The principle of two-dimensionally projected space-composition, as *fixed* by a groundplan, will be replaced by exact *calculation of the construction*, a calculation which must transfer the carrying capacity to the simplest but sturdiest points of support. Euclidean mathematics will no longer serve this purpose; yet by using Non-Euclidean calculations in four dimensions, this calculation can be accomplished quite easily [Van Doesburg 1924: 144].

The need for non-Euclidean geometry in architectural design points to the fact that space and time are involved. Van Doesburg himself made this clear:

> The new architecture calculates not only with space but also with time as an architectural value. The unity of space and time will give architectural form a new and completely plastic aspect, that is, a four-dimensional, plastic space-time aspect [Van Doesburg 1924: 144].

Van Doesburg unfortunately supplied vague descriptions on how these principles could be translated into concrete terms. Despite noting how simply one can conceive an architectural model using Non-Euclidean geometry, there exist no examples of this technique in Van Doesburg's own sketches or descriptions in his writings. We do know though, that the use of colour was important in creating the four-dimensional aspect of Van Doesburg's new conception of architecture: "The new architecture employs colour organically as a direct means of expression of relationships in space and time" [Van Doesburg 1924: 145]. But again, as was the case with the groundplan, little is said as to how colour functions in this regard.

However, a unique feature of Van Doesburg's design is that there is never one fixed point from which one can define the whole of the structure. Every vantage point provides a unique view that is never repeated. In other words, there is no defining moment, no fixed or absolute point, and thus Van Doesburg achieves an inventive type of completely relativistic, Dadaist type of architecture. It embodies a notion we will encounter with Lissitzky, that every point in space is related to a unique moment in time.

The exterior appearance is complemented by the interior, for which Van Doesburg proposed the use of moving walls/partitions that would allow for a variety of interior configurations. This is an idea that would be employed in what is arguably the only De Stijl structure, the Schröder House, designed by Gerrit Rietveld with the help of Truus Schröder in 1942-25. Rietveld was a furniture designer who joined De Stijl in 1919. He helped Van Doesburg and Van Eesteren with the Rosenberg commission, designing the models, and his work with them contributed to his creation of some unique pieces of furniture, the Schröder Table and Berlin Chair (1923), which followed the principles being outlined by Van Doesburg. Obviously, Rietveld embraced Van Doesburg's redefining of architecture, since the Schröder House would follow a number of the suggestions laid out by Van Doesburg.

Van Doesburg's architectural theories and designs must have inspired Lissitzky. In "A. and Pangeometry," Lissitzky set out the possible means for creating imaginary space in art, beyond the confines of painting:

> ... we know that a material point can form a line; for example: a glowing coal while moving leaves the impression of a luminous line. The movement of a material line produces the impression of an area of a body. There you have but an intimation of how one can build a material object by means of elementary bodies, in such a way that while it is motionless it forms a unity in our three-dimensional space, and when set in motion it generates an entirely new object, that is to say, a new expression of space, which is there for as long as the movement lasts and is therefore imaginary. ... Motion is incorporated ... as an ingredient in the total complex of the elements which are to build the new bodies [Lissitzky 1968: 352-353].

This entailed the use of objects which function on the basis of rapid rotation or vibration. Lissitzky illustrated an example of it in his text "A. and Pangeometry" and also mentioned a work by Naum Gabo which, in Lissitzky's words, "stylized the pendulum-movement of a metrodome".[5] For Lissitzky, the space generated by the rapid movement of an object is imaginary for the simple reason that it only exists "as long as the movement lasts." Once the movement ceases, the object returns to its original state as part of our three-dimensional reality. But such works were nothing more than illustrations of imaginary space, which explains why Lissitzky did not experiment with kinetic sculpture. He wanted to move one step further by creating a work in which one could physically experience the unity of time/movement and space.

Lissitzky had constructed an actual physical space based on his Proun imagery before 1925, the *Proun Room* (1923) (fig. 14), in part encouraged by Van Doesburg's own work in architecture. Upon its walls were affixed three-dimensional recreations of Proun paintings. The objective was to generate a living space that encouraged one to walk within it. Thus, movement was a component of the work. That the theme was to somehow build an environment in which one could experience the unity of time and space is suggested by the fact that Lissitzky had planned to adapt the composition of *Proun G7* for the ceiling of the *Proun Room*. This theme was radically reformulated in the Dresden and Hanover Exhibition Rooms of 1926 and 1928 respectively.

In the Dresden *Room for Constructivist Art* (fig. 15), Lissitzky placed thin vertical strips perpendicular to the wall, strips which were seven centimeters wide and placed at seven centimeter intervals. The wall itself was painted gray and the strips were painted white on one side, black on the other and the ends gray. The result was that as one moved within the room, its color shifted gradually from white to gray to black. Added to this was the hanging of the pictures in the room on moving panels, which allowed the viewer to physically participate in the transformation of the room. The Hanover room, *The Abstract Cabinet*, (fig. 16) repeated these devices with some modifications and additions. The movable paintings were complemented by rotating showcases for the sculpture. Instead of vertical strips, Lissitzky used thinner triangular ones (three centimeters wide at their base), spaced at smaller intervals (two centimeters apart); the sides of the strips were once again painted white on one side, black on the other and gray at the tip. The modification of the strips made the transformation in wall color more gradual, intensifying the effect presented in Dresden.

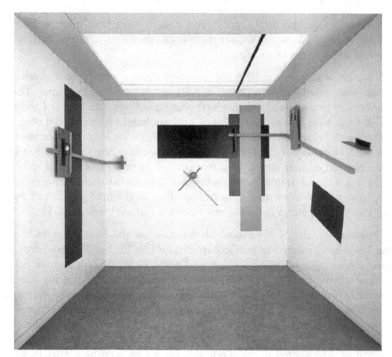

Fig. 14. El Lissitzky, *Proun Room*, 1923, reconstruction, 300.0 x 300.0 x 260.0 cm

Fig. 15. El Lissitzky, *Room for Constructivist Art*, Dresden, 1926, 6 x 6 m

Fig. 16. El Lissitzky, *The Abstract Cabinet*, Hanover, 1928, 300.0 x 427.0 x 549.0 cm

As spectators at a 1991 retrospective in Paris of Lissitzky's work were able to witness in a reconstruction of the Hanover room, the effect created by this exhibition space on the viewer was remarkable. It was best experienced by visualizing the appearance of the room just before entering, and then proceeding into it with one's head down; once inside, the immediate impression upon raising one's head was a sense of disorientation, caused by the feeling that the space you are in is not the same space you saw just before entering. As one moved within the room, its appearance changed continuously in a mesmerizing way. This accomplished Lissitzky's goal of uniting time and space, where for each point in time, defined as movement, there was a unique spatial configuration, i.e., visual appearance of the room or "room-space," as Lissitzky called it. The Hanover room represented Lissitzky's most ingenious and successful aesthetic embodiment of Minkowski's statement that "Nobody has ever noticed a place except at a time, or a time except at a place." This clearly echoes what happens with the Rosenberg commission designs.

The failure of the Rosenberg commission to materialize resulted in Van Doesburg's return to painting. But Van Doesburg's experiments with architecture were not abandoned; they would find their way into his paintings, resulting in some rather significant and controversial changes, most notably Mondrian's departure from De Stijl. Mondrian could live with Van Doesburg's interest in time in art when it concerned itself with architecture, since this was by definition a materialistic art form. But once Van Doesburg began to boldly incorporate these ideas into painting, Mondrian could no longer endure the corruption of his own Neo-Plastic ideals.

Van Doesburg claimed by 1926 that his paintings were a "plastic intuition, controlled by a scientific idea, which is needed by the new man" [Van Doesburg 1926a: 155]. This debt to the Theory of Relativity was further spelled out in "Painting and Plastic art", when in defining "Elementarism", the name Van Doesburg gave to his new art, he wrote:

> *Elementarism* is the equivalent of relativity, of the latest discoveries about matter and of phenomenological definitions concerning the unlimited, yet latent, omnipotence of human intelligence. In contrast to religious dogmatists [an obvious slur against Mondrian], the Elementarist considers life only as 'a perpetual transformation' ... [Van Doesburg 1926b: 160].

He added that

> Elementarism is preparing for the realization of elementary counter-plastic form, and we must first destroy the use of the static axis in contempt for

the Euclidean view of life (which relates to the static point) [Van Doesburg 1926b: 160].

Elementarism thus summed up all of Van Doesburg's experiments in art dating back to around 1918 and which had essentially remained in the background until 1926. Strangely enough, Van Doesburg reconciled with Mondrian shortly before his death in 1931, and re-embraced his mentor's positions on art.

Lissitzky's interest in science and its role in his art continued but in a somewhat more muted form necessitated by the rise of Stalinism. For Lissitzky the Theory of Relativity supplied the most fundamental reformulation of reality occurring in his time and, more importantly, was part of a broader cultural change, where:

> ... the confines of expertise have been blown to bits. Methods which were once employed in a particular branch of art, knowledge, science, philosophy, are now being transferred into other areas. This is happening, for example, to the four coordinates of Minkowski's world ... [Lissitzky 1976: 60].

Lissitzky's art was extensively nourished by "Minkowski's world" and, for Lissitzky, the future rested with a better understanding of science. It is certainly for this reason that Lissitzky began planning and designing a mathematics book for children in 1928.

Acknowledgments

I must thank Dawn Ades, Linda Dalrymple Henderson, Kim Williams, Gerard Curtis, and Danika Irvine for their help at various stages of the research and writing; my usual nod to Karen, Robbie and Patrick, who patiently put up with my obsessions; and, for their invaluable financial support, many thanks to the Social Sciences and Humanities Research Council of Canada. However, as appreciative as I am for all this help, this article is dedicated to my parents who passed away a few years ago and whose absence has created an immeasurable void in my life and those who knew them; they always reminded me that my first crib was a gift from the Physics Department at Queen's University

Notes

1. [Minkowski 1908: 84, fig. 2]. This collection of essays was originally published in Germany in 1920 and was already in its third edition by 1923.
2. [Bonset 1920: 114-15]. Blotkamp translates the last line as "the pell mell of nothingness and being"; cf. [Blotkamp 1986: 30].
3. Most of the effects described pertaining to *Proun 99* are found in the Proun illustrated in fig. 9, with the exception of the two-dimensional strip.
4. [Weyl 1922: 169-177]. English readers of texts on Relativity will probably be more familiar with A.S. Eddington's "Absolute Future, Absolute Past and Here-Now hourglass model". This model is essentially the same as that published by Weyl nine years earlier. The similarity of these models is due to the fact they are both based on Hermann Minkowski's seminal work on Relativity; see [Eddington 1928: 41-50].
5. [Lissitzky 1968: 352]. The Gabo work in question is the *Kinetic Construction* (1919-20: The Tate Gallery, London).

References

BALJEU, Joost. 1974. *Theo van Doesburg*. New York: Macmillan.
BIRNHOLZ, Alan C. 1983. El Lissitzky. Ph.D. dissertation, Yale University, New Haven.
BLOTKAMP, Carel. 1986. Theo van Doesburg. In: Carel Blotkamp et. al., *De Stijl: The Formative Years, 1917-1922*, Cambridge, MA: M.I.T. Press.
BONSET, I. K. 1920. X-Images. In: Joost Baljeu, *Theo van Doesburg*, New York: Macmillan, 1974.

BONSET, I. K. 1923. What is dada???????. In: Joost Baljeu, *Theo van Doesburg*, New York: Macmillan, 1974.

EDDINGTON, A.S. 1928. *The Nature of the Physical World*. Cambridge: Cambridge University Press.

EL LISSITZKY. 1968. A. and Pangeometry (1925). In Sophie Lisstzky-Küppers, *El Lissitzky. Life, Letters, Texts*. London: Thames and Hudson.

———. 1976. Proun (1920-21). In *El Lissitzky*. Cologne: Galerie Gmurzynska.

HATCH, John G. 1995. Nature's Laws and the Changing Image of Reality in Art and Physics: A Study of the Impact of Modern Physics on the Visual Arts, 1910-1940. Ph.D. diss., University of Essex.

HIROSIGE, Tetu. 1976. The Ether Problem, the Mechanistic World View and the Origins of the Theory of Relativity. *Historical Studies in the Physical Sciences* 7: 3-82.

HOLTZMAN, Henry and Martin S. JAMES, eds. 1986. *The New Art – The New Life: The Collected Writings of Piet Mondrian*. Boston: G.K. Hall & Co.

JAMES, Martin. 1957. The Realism behind Mondrian's Geometry. *Art News* LVI (Dec. 1957): 34-37 & 59-61.

MONDRIAN, Piet. 1922. The Realization of Neo-Plasticism in the Distant Future and in Architecture Today. In: *The New Art – The New Life: The Collected Writings of Piet Mondrian*, Henry Holtzman and Martin S. James, eds., Boston: G.K. Hall & Co., 1986.

MINKOWSKI, H. 1908. Space and Time. In: A. Einstein, H.A. Lorentz, H. Minkowski and H. Weyl, *The Principle of Relativity: A Collection of Original Memoirs on the Special and General Theory of Relativity*. Rpt. New York: Dover, 1952.

NISBET, Peter, ed. 1988. *El Lissitzky, 1890-1941*. Hanover, FRG and Cambridge, MA.: Busch-Reisinger Museum and Harvard University Art Museums.

PERLOFF, Nancy Lynn and Brian REED, eds. 2003. *Situating El Lissitzky: Vitebsk, Berlin, Moscow*. Los Angeles, CA: Getty Research Institute.

TUPITSYN, Margarita et. al. 1999. *El Lissitzky: Beyond the Abstract Cabinet: Photography, Design, Collaboration*. New Haven: Yale University Press.

VAN DOESBURG, Theo. 1922. The Will to Style: The New form Expression of Life, Art and Technology. In: Joost Baljeu, *Theo van Doesburg*, New York: Macmillan, 1974.

———. 1924. Towards Plastic Architecture. In: Joost Baljeu, *Theo van Doesburg*, New York: Macmillan, 1974.

———. 1926a. Painting: From Composition towards Counter-Composition. In: Joost Baljeu, *Theo van Doesburg*, New York: Macmillan, 1974.

———. 1926b. Painting and Plastic Art: On Counter Composition and Counter-Plastic Elementarism (A Manifesto Fragment). In: Joost Baljeu, *Theo van Doesburg*, New York: Macmillan, 1974.

———. 1927. Painting and plastic art: Elementarism. In: Joost Baljeu, *Theo van Doesburg*, New York: Macmillan, 1974.

WEYL, Hermann. 1922. *Space, Time, Matter*. (Rpt. New York: Dover Publications).

About the author

Dr. John G. Hatch is an Associate Professor in the Department of Visual Arts at The University of Western Ontario, where he has taught modern art history and theory since 1994, and is currently the Associate Dean (Academic) for the Faculty of Arts and Humanities. He received his Ph.D. in Art History and Theory from the University of Essex in 1995. His research interests are diverse and have resulted in publications on the work of Max Ernst, Francis Bacon, Cindy Sherman, Robert Rauschenberg, Frantisek Kupka, to name a few. He has a particular fascination for the influence of the physical sciences on art and architecture, ranging from examining the use of Keplerian cosmology in the seventeenth-century buildings of Francesco Borromini to looking at the impact of entropy on the earthworks and writings of the American artist Robert Smithson.

Book Review

Samuel Y. Edgerton

The Mirror, the Window, and the Telescope: How Renaissance Linear Perspective Changed Our Vision of the Universe

Ithaca, NY : Cornell University Press, 2009

Reviewed by the author

940 Hancock Road
Williamstown, MA 01267 USA
Samuel.Y.Edgerton@williams.edu

Keywords : Samuel Edgerton, perspective, Galileo, Brunelleschi, Alberti, Fra Angelico, Renaissance art

My new book, *The Mirror, the Window, and the Telescope: How Renaissance Linear Perspective Changed Our Vision of the Universe*, both revises and extends my earlier, but now out of print (in English), *Renaissance Rediscovery of Linear Perspective* (New York, 1975/6). Furthermore, it also draws from my subsequent book, *The Heritage of Giotto's Geometry: Art and Science on the Eve of the Scientific Revolution*, (Cornell University Press, Ithaca, NY, 1991/3). The subject once again has to do with the inception of geometric linear perspective in the pictorial arts during the early Renaissance. I begin by tracing the reasons why this mathematical method was conceived only in Western Christendom (no other culture in the world had ever invented it independently before). The failure of the Crusades, the loss of Jerusalem, and schismatic divisions within the Church itself had badly weakened the Faith by the late Middle Ages, and many felt that religious imagery needed to be refreshed in order to help rekindle Christian fervor. Some application of the revived ancient science of Euclidian geometry might be the answer.

While there had been a slow, empirical movement in this direction since the fourteenth century, the first application of optics – the geometric sub-science of vision and light – to the painting of pictures is credited to Filippo Brunelleschi of Florence, Italy. Around 1425 he demonstrated how a familiar building in the city, the Baptistery beside the Cathedral, could be painted in a picture exactly as it appeared in a mirror reflection. Furthermore, the test he devised to prove the "realism" of his unique painting was that the viewer should hold it in one hand and peak through a hole drilled though the backside, with the painted-side reflected in a mirror held before it in the other hand. In other words, the viewer was to judge the picture's verity not by comparing the image to the actual building, but only by comparing it to its own mirror reflection. I argue that the reason for this peculiar test had more to do with medieval religious assumptions than any prescient anticipation of modern science. In fact, what Brunelleschi hoped to reveal was the very process by which God created the universe at Genesis when he projected it into the void as mirrored from his divine mind's eye. Brunelleschi believed, as did all good Christians during the Middle-Ages, that living mortals are permitted to view the

---

DOI 10.1007/s00004-010-0020-x; *published online* 9 February 2010
© 2011 Kim Williams Books, Turin

world only as a pale reflection of true "reality" in heaven, just as Saint Paul stated in his Epistle to the Corinthians I, 13:12, "At present we see indistinctly, as in a mirror, but then [in heaven] face to face."

Brunelleschi quickly communicated the principles of this new perspective to his artist friends, Donatello, Masaccio, Masolino, and Fra Angelico. However, there was still a problem concerning its acceptance . Some painters, like Fra Angelico, began to worry that geometry by its very definition referred only to "earth measure." While linear perspective was the optically correct means for representing the material world, it might be incorrect and even profane when applied to depictions of ineffable heaven and sacred mysteries.

Thus in 1435/6, the humanist scholar and fellow Florentine Leon Battista Alberti decided to rectify the matter in a treatise called *De Pictura,* On Painting. First, he codified Brunelleschi's perspective rules into a simple formula that even mathematically disadvantaged artists could understand, and next he replaced Brunelleschi's mirror with a gridded window as the ultimate test for pictorial "reality." In other words, by looking through a window the geometric beauties of this world were revealed without the theological implications of a mirror reflection. Moreover, Alberti stressed that the proper subjects to be depicted in his perspective window need not necessarily be religious but were preferably the historical events of ancient Greece and Rome. Paintings of these secular but still hallowed classical stories should then serve as didactic models for the moral edification of living viewers. The proper purpose of geometric linear perspective was therefore to make it possible for artists to represent nature not as divine mystery but idealized as worldly perfection.

Alberti's book, subsequently printed in 1540, spread the new concept everywhere in Italy and transalpine Europe. Indeed, "Alberti's window" became the trademark of High Renaissance art, especially by way of Raphael's proliferating influence. Raphael was perhaps the most effective sixteenth-century promulgator of Alberti's classical vision. While he and his followers still painted beautiful pictures of religious subject matter, their images of holy personages were posed more often to look like Olympian gods than ascetic Christian saints. Even their representations of supposedly ethereal heaven appeared more and more like manicured earthly landscapes, In fact, Renaissance artists in general tended to frame heavenly space according to the same Albertian rules, as if it were seamlessly contiguous with mundane space in spite of orthodox Christian doctrine.

Finally, I show how these new applications of linear perspective even influenced Galileo Galilei in the early seventeenth century. In fact the new telescope that the Florentine physicist and astronomer constructed followed the same optical principles that had been the basis of Brunelleschi's and Alberti's pictorial experiments. He most famously pointed it at the skies to observe the moon and other planets. Being also an expert in perspective and *chiaroscuro* (light and dark) drawing, he was able for the first time ever to understand that the heretofore "strange spottedness" of the moon was actually caused by the shades and shadows of high mountains and deep valleys. He even drew his own perspective pictures of the rugged lunar surface which were published in his sensational book, *Sidereus Nuncius* (Starry Messenger) printed in 1610, demonstrating beyond any doubt that the moon, the immaculate pearl in Dante's poetic ascent to Paradise, was hardly the smooth sphere exuding heavenly perfection as always assumed by the Church.

Moreover, Galileo's telescope, called at the time a "perspective tube," verified what Renaissance artists were already depicting. What it indeed revealed was that the earth was not necessarily a mirror reflection of the heavens as Brunelleschi's mirror had advocated, but just the other way around.

This book updates my earlier books on perspective because I am adding new evidence concerning the intellectual ambience of Florence, Italy, during the early fifteenth century. I also submit new evidence derived from examination of works of art from the years between 1413 and 1436 regarding the date and methods employed by the first Florentine artists who adapted the rules of optical science to their traditional, empirical understanding of visualized nature.

The book is also a sequel because of my concern that the subject of perspective in the arts has fallen victim to a wave of art criticism which no longer considers it a positive idea; that perspective has instead actually inhibited innate artistic expression, even becoming an imperialist means to colonize the cultures of non-Western societies. Indeed, there is a tendency nowadays to downgrade the importance of perspective as merely a brief side-track in the evolution of world art. Thus, I try to re-connect the advent of perspective to its roots in the intense religious and moral preoccupations of the European late Middle Ages, to the "period eye" of the Renaissance in Michael Baxandall's famous phrase. Whatever one may say about the eventual use or misuse of geometric perspective as a tool of Western political power, it was surely conceived in the early fifteenth-century as a very medieval Christian solution to a very medieval Christian problem. It must be understood in the context of the strongly held spiritual beliefs and assumptions of still devout Christians who longed for painted and sculpted images that could arouse the feeling of divine presence and reinforce their faith that God and his saints were still immanent in their daily lives.

I have found strong testimony to this in the preaching of Fra Antonino Pierozzi (1389-1459), Dominican prior of the San Marco convent (when Fra Angelico painted there) who then became the influential Archbishop of Florence. While the writings of Antonino (eventually canonized as St Antonine) have received some attention from modern scholars, I believe I am the first to single out his considerable views on optics, indicating just how au courant this subject was in fifteenth-century Florence. I speculate with good circumstantial reason that among the avid listeners to Antonino's popular sermons was none other than Filippo Brunelleschi.

In the same light, I offer new analyses and interpretations of art by Masaccio, Masolino, Donatello and Fra Angelico before and after 1425 that surely indicate that Brunelleschi must have performed his famous perspective demonstrations on or about that year. Also I will show that these artists exhibited most of the same perspective principles and short-cuts in their own works between 1425 and 1435 that Leon Battista Alberti described in his *De Pictura* after 1435, thus indicating further that his own famous perspective system was basically a verbal codification of what Brunelleschi had already achieved pictorially ten years before.

About the author

Samuel Y. Edgerton is Amos Lawrence Professor of Art History Emeritus at Williams College, where he taught for twenty-seven years, and at Boston University for sixteen years before that. During those four plus decades his scholarly interests have ranged from studying the arts of medieval and Renaissance Europe to the arts of pre- and post-conquest America. However, the single thread that unites the seemingly diverse subjects of his books is his desire to reveal how the history of art interacted with the ideologies and social institutions of these diverse cultures, such as the way art was deployed in the service of the criminal justice system in still medieval Florence, or the way Spanish missionaries used the arts to help convert the indigenous peoples of sixteenth-century Mexico. Edgerton's latest work again traces the advent of artistic linear perspective, how it was originally conceived to reinforce the devotional power of Christian pictures; how it then became the universal trademark of Renaissance artistic "realism"; and finally how perspectival art allowed Galileo Galilei to "see" scientifically for the first time the true form of our heavenly universe.

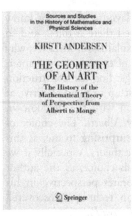

Book Review

Kirsti Andersen

The Geometry of an Art. The History of Perspective from Alberti to Monge

New York : Springer, 2007

Reviewed by João Pedro Xavier

Faculdade de Arquitectura da Universidade do Porto (FAUP)
Via Panorâmica S/N
4150-755 Porto PORTUGAL
jpx@arq.up.pt

Keywords: Kirsti Andersen, perspective, perspective geometry, Leon Battista Alberti, Gaspard Monge, Guidobaldo del Monte

Kirsti Andersen's book, *The Geometry of an Art. The History of Perspective from Alberti to Monge*, will surely become one of the fundamental references concerning the mathematical development of *perspectiva artificialis*, from the first steps, found in *De Pictura* (1435) up to its integration in the last edition of *Géométrie Descriptive* (1820).

Given the wide scope of this study and the author's methodology, based on an extensive and careful analysis of "more than two hundred books, booklets, and pamphlets on perspective" [Introduction, xxi], *The Geometry of an Art* is a monumental work, the work of a lifetime, one could say. The author's lifetime has been filled with accurate investigations on perspective that has already given rise to some of the most important titles about its history regarding the work of Piero della Francesca, Stevin, Desargues and Brook Taylor, among others.

This is actually a book of history of science dealing with a specific branch of geometry, that is, perspective. Although perspective became the geometry of an art, because its history is related to the conquest of exactitude in the representation of space from an identifiable point of projection (the artist's viewpoint), the main scope of this book is to narrate the process leading towards the definition of a mathematical theory of perspective:

> My primary sources do not give an adequate background for discussing thoroughly the highly pertinent question of the actual use of perspective in paintings, architectural illustrations, and other drawings [Introduction, xxiii].

In a certain way we needed a comprehensive work that systematically treats the scientific aspects of perspective because these were so often neglected and biased in favour of dubious interpretations of its significance. That is why the author is very careful about her sources and only studies what come from printed materials containing drawings and explanations, such as perspective textbooks or other treatises where the subject is developed. This seems to me a wise approach, because even if we look at paintings artifically with the aid of x-rays or infrared to find perspective layouts we cannot know for sure which perspective construction underlies them (if indeed there is one... and so many have been "found" that could never exist!). This attitude explains, for instance, why the study begins with Alberti rather than Brunelleschi. In spite of this, the first chapter, "The birth of perspective", discusses some of the hypotheses regarding the construction of Brunelleschi's famous *tavolette* but, as it is impossible (at least up to now) to arrive at any certainties, the author prefers to begin the main discussion with the first known description of a perspective construction, the one provided by Alberti. The choice of ending the discussion with Gaspard Monge, seems a little bit surprising to me, as this French geometer did not introduce any novelty into perspective theory, especially if we compare him to Johann Heinrich Lambert. The reason for this choice, as the author explains, is the idea of *closing a cycle*, since Monge incorporates perspective in his *Géométrie Descriptive* by returning to the plan and elevation technique discovered earlier, and for the first time, in Piero della Francesca's *De Perspectiva Pingendi*.

Also impressive is the wide range of countries and nationalities covered: Italy, Germany, France, Holland and England, areas of Europe which actually were the main stages for the development of perspective. No Spanish or Portuguese contributions were studied, which I presume was due to difficulties with the language. The author writes, "I am confident that my material is so comprehensive that adding further publications would not change my conclusions in any significant way" [Introduction, xxii]. I believe I can confirm that the Iberian works cast no doubt on the author's conclusions.

In fact, as far as both Spain and Portugal are concerned, what matters is the way perspective constructions are either assimilated (with much misinterpretation as in many other places), or simply refused in favour of alternative techniques, possibly inspired by instruments for taking angular measures. This seems to be the case of the so-called "Spanish Renaissance angular perspective" (studied by Lino Cabezas[1]) developed and presented by Rodrigo Gil de Hontañon[2] (ca. 1500-1577) and Hernán Ruiz, el joven[3] (1514?-1569) around 1560, which is an unorthodox and alternative perspective construction similar to the one that was to appear later in Gianfrancesco Costa's *Elementi di prospettiva per uso degli architetti e pittori* (1747), applied in a different context and pursuing other purposes. This perspective construction is treated by Andersen in the section "A Special approach to Perspective – Costa" [VIII.6, 394].

In order make the investigation coherent and evaluate the relative importance of the protagonists of this story Kirsti Andersen conducts her research using a query composed of the following questions:

> How does the work of the author relate to other literature on perspective?
> Which perspective constructions did he chose? How did he describe them
> and, in particular, to what extent was his presentation based on a
> geometrical background? [Introduction, xxii].

The data concerning each author's knowledge of previous and relevant contributions, the insights added to improve the quality of a specific perspective construction or to provide a new one, the capacity to arrive at complete explanations enlightened by a theoretical mathematical background, led Andersen to select a group of protagonists comprising Guidobaldo del Monte, Simon Stevin, Willem 'sGravesande, Brook Taylor and finally Lambert, to whom the creation of perspective geometry is credited.

Let me summarize, using here and there a few Andersen's own words, the main achievements of these most significant figures in the history of perspective.

Guidobaldo is clearly the geometer who marks the end of a cycle and the beginning of a new one. In fact, although he is still concerned with the problem of the section of the visual pyramid, the Albertian *intersecazione*, he turns the interest to what happens in the picture plane as a result of that intersection, which means, in other words, that he became concerned with perspective itself. His main contribution is the definition of the *punctum concursus* and the generalization of that concept for any direction. This is the vanishing point theorem, to use Taylor's updated language, which is directly related to what Andersen calls the *Main Theorem of Perspective*:

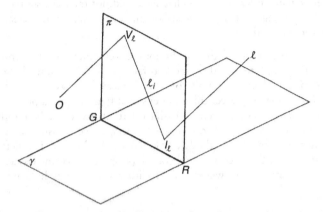

The Main Theorem of Perspective (FIGURE VI.5) The image l_i of any line l that is not parallel to the picture plane is determined by the vanishing point V_l and the intersection I_l
[VI.2, 245-246]

Although it is possible to detect in the first ideas about the *punctum concursus* some traces of an operative practice, the fact is that its full development puts Guidobaldo in the core of the theory of perspective. This is very interesting, because this scientific iter led him to join the northern empiric tradition, begun with Viator and clearly visible in Vredeman de Vries, characterized by the idea of describing space directly on the picture plane from visual observation.[4] In the end, Lambert's notable conquest was having revealed the mathematical key to this procedure and achieving the full development of a language of perspective geometry although, paradoxically, this language could only be spoken by a few, and would remain almost inaccessible to practitioners (see [XIV.4, 720 "The Usefulness of the Theory of Perspective"]).

According to Andersen, Stevin was clearly a follower of Guidobaldo and continued the research deep inside the theory of perspective. Feeling the need to build the foundations of a new geometry he tried to establish a set of definitions and postulates (axioms) in a Euclidean manner, from which he deduced six theorems, of which the third "is in effect Guidobaldo's important vanishing point theorem, stated in complete generality" [VI.7, 271]. His influence, and possibly also the insights of van Schooten, reached 's Gravesande, who pushed it further "as he appreciated and exploited its potential more strongly than his predecessors, for instance by focussing upon a very elegant and simple visual ray construction" [XIV.3, 717]. Then 's Gravesande's mathematical understanding of perspective inspired Taylor, who "provided perspective with a new mathematical life, among other things by introducing and applying the general concept of a vanishing line" [XIV.3, 717].

Before returning to Lambert we shall remark the surprising absence of Desargues from Andersen's list of protagonists, particularly if we are talking about the concepts of vanishing points and lines, or in the terminology of the French architect, the points and lines at infinity. Desargues, with his *Brouillon Project...*, can be thought of as the grandfather of Poncelet's projective geometry. However, as the author convincingly argues, the fact is that Desargues's general method, his *manière universelle* applied to perspective, does not reflect the use of such concepts, although it is hard to prove that the first steps in the field of projective geometry were unknown and could not have inspired Lambert. Here Kirsti Andersen reveals that she could not trace the sources for Lambert's mathematical knowledge of perspective although, as she remarks, it is clear that he could not have reinvented the theory:

> His work was part of the continuous development of the theory of perspective, and brought this theory as far as it could presumably be taken as an independent discipline. By this I mean that as for the specific question of how to project three-dimensional figures upon a plane surface, no important question seemed to have left unanswered. Nor did there seem to be any way of carrying out perspective constructions more elegantly than by the methods advocated by Taylor and Lambert. As a consequence of this success, the theory of perspective became less attractive for mathematicians, to whom a field with no loose ends hold no appeal [XIV.3, 716].

And so, for mathematicians, projective geometry became more attractive than perspective. But as late as 1775 perspective could still seduce a devotee of Brook Taylor, Thomas Malton, to write, "of all the Mathematical Sciences, the study of perspective is perhaps the most entertaining" [XIV.5, 721].

This is the final quotation chosen by Kirsti Andersen to close her remarkable study on perspective, which is indeed, the Geometry of an Art.

Notes

1. See Lino Cabezas, Tratadistas y tratados españoles de perspectiva, desde los orígenes hasta la Geometría Descriptiva de Gaspard Monge (Ph.D. Thesis, Universidad de Barcelona, 1985), p. 181-209.
2. Rodrigo Gil's "angular perspective" is known through the transcription made by Simón García in *Compendio de Architectura y simetria de los templos conforme a la medida del cuerpo humano. Con algunas demostraciones de geometría* (1681). Manuscript 8884, Madrid, Biblioteca Nacional.
3. The *Libro de Arquitectura* (c. 1560) by Hernán Ruiz el Joven is kept in the rare books section of the Library of the Scuela Superior de Arquitectura de Madrid. There is a critical edition of this book: Pedro Navascués Palacio, *El Libro de Arquitectura de Hernán Ruiz el Joven*. Madrid: Xarait, 1974.
4. Svetlana Alpers, "'Ut pictura, ita Visio': Kepler's Model of the Eye and the Nature of Picturing in the North" in *The Art of Describing. Dutch Art in the Seventeenth Century*. Chicago: The University of Chicago Press, 1983, p. 26-71.

About the reviewer

João Pedro Xavier is an architect and geometry teacher in the Faculty of Architecture of the University of Porto (FAUP) where he was graduated in 1985 and received a Ph.D. in Architecture in 2005. He worked in Álvaro Siza's office from 1986 to 1999 and, at the same time, he established his own practice as an architect (http://www.jpx-arq.com). Xavier has always been interested in the relationship between architecture and geometry, especially perspective. He is the author of *Perspectiva, perspectiva acelerada e contraperspectiva* (FAUP Publicações, 1997) and *Sobre as origens da perspectiva em Portugal* (FAUP Publicações, 2006). He published several works and papers on the subject, presented conferences and lectures and taught courses to high school teachers.

Book Review

Javier Navarro de Zuvillaga

Forma y Representación. Un Análisis Geométrico

Madrid : Ediciones Akal, 2008

Reviewed by Tomás García-Salgado

National Autonomous University of Mexico
Palacio de Versalles 200
Col. Lomas Reforma
MÉXICO D. F. C.P. 11930
tgsalgado@perspectivegeometry.com

Keywords : Javier Navarro
de Zuvillaga, form,
representation, perspective,
geometry

If I had to summarize what this book is about in a few words, I would say that it is about the geometry of the form and its representation in perspective (as the title says, Form and Representation). It is an interesting and enlightened book comprising five chapters and an appendix, which turns out to be another chapter written by Juan Chamorro Sánchez. The thematic of the book embraces an ample number of topics over its five chapters. Sometimes the topics seem to be repeated, but within a different context, as for instance "scale and proportion" and "representation and scale"; a careful reading of the index makes the contexts clear. Because of the extent of the topics the book is recommended as a general reading on geometry and perspective.

In Chapter 1, the author begins with the notions of human body, space, and geometry. Here, the course of the ideas and illustrations are discussed generally, describing the principles of scale and proportions in nature, architecture, astronomy, and art. It continues with the notions of movement, and visual perception of space and of objects. In a simple manner the author explains how a line, a plane, and a sphere are generated by respectively translating points and lines, and rotating a radius. At this point, the author remarks on the origin of projective geometry during the Renaissance, when a plane intersecting the cone of vision captured the image of the observer. A fundamental principle of perspective first established by Alberti in his treatise *Della Pittura*, in which the intersecting plane of the visual pyramid (or cone of vision) is called a "*finestra*" (window).

Form and Representation is the theme of Chapter 2. It begins with a definition of "form" and its classification in three kinds: natural form, geometric form, and created form. These three species relate to "representation" in many ways which in turn generate systems and models of representation. In this section, the illustrations vary in gender and epoch, thus paintings, analytic figures, geometric outlining, or scientific images were selected to enhance the author's idea about systems of representation. Of course, I could not agree more with the author's choice of Piero's *Città Ideale* as one of the paramount examples of created form. Languages of representation imply the knowledge of perspective, colors, and models. The language of line as a perspective system is behind Velázquez's *Las Meninas*, as the author points out. Instead, in architecture, or scenography, the language of representation used, whether 2D or 3D, would have to

DOI 10.1007/s00004-010-0022-8; *published online* 10 March 2010
© 2011 Kim Williams Books, Turin

produce tangible things in the end. For example, a scenography design in 3D with a perspectival effect is literally intended to construct an object in perspective: a perspective-object to be seen in perspective once it is settled in place, such as the seven streets of the city of Thebes in Palladio's Teatro Olimpico.

Chapter 3 is devoted to the Geometry of the Plane. Here the author introduces the following notions: geometrical plane, measure systems, scales, and proportions. In particular, the construction of regular figures on a plane, and how they are used to conform architectural elements, or serve as layouts in art, decoration, and graphic design, is presented by means of well-selected examples. While a plane is theoretically infinite in geometry, it is finite for art and design. However, a real plane, or a material one, can sometimes be ambiguous, such as the Alhambra mosaics that can virtually expand indefinitely, despite of their being confined within walls. Next, it continues with the notions of prisms, polygons, networks, and how they are used in architectural design to modulate the walls, floors, ceiling, and windows. In art, the concept of the infinite within a finite plane is illustrated by Escher's works.

The Geometry of the Space is the theme of Chapter 4. Here the author introduces the idea of "plastic space", a "space" based in the real space feeding the imagination of the artists in the creation of plastic forms. In other cases, a geometrical form turns out to be a built structure, as the equilateral pyramid of the Louvre, with its base opened and faces thoroughly modulated by thinner pyramids. Partially visible, the dodecahedron structure housing Dalí's *Last Supper* is a remarkable example of a "plastic space" created in painting. Polyhedra are the main topic along this chapter in regard to its application in art and architecture. In addition, some geometrical operations of polyhedra, such as axial rotation, symmetry, projective views, sectional views, combination, inscription, circumscription, duality, and intersection, are explained. To complete this topic, the author gives special attention to describing the geometry of the sphere, and so of radial surfaces, warped surfaces, surfaces of revolution, modular networks, and the conics.

The title of Chapter 5 is Systematic Representation. Here, some of the topics discussed in previous chapters come together under the idea of "representation". Switching from 2D to 3D, and vice versa, from 3D to 2D, the author explains how central and parallel projection sometimes produce ambiguities or impossible figures. When one sees a figure representing a square, a triangle and a sphere, it also represents a cube, a pyramid, and sphere, depending on our level of perception. For instance, when looks at a rhomboid, it takes a little while to realize that it can also be seen as a cube in perspective. The so-called projections – orthogonal, perspective, dihedral, axonometric, plane-rotated, with their corresponding variants –, are at the core of what the author calls "systematic methods of representation." In this chapter, the author expands on the topic of perspective, illustrating different geometrical methods of representation, referring Piero's *De Prospectiva Pingendi*, Bosse's *Traité des Pratiques Géométrales*, Monge's *Géométrie Descriptive*, Dubreuil's interpretation of Leonardo's window, Dürer's windows, Saenredam's view of the Great Church of Harlem, Peruzzi's Villa Farnesina, and some examples of anamorphic perspective. It then continues with a theory of shadows in perspective that includes a sphere in dihedral projection with its own degraded shadows, while another example shows a sphere casting its shadow on a plane. This chapter concludes with diagrammatic representation, a topic not often discussed.

The appendix, entitled Virtual Space, is essentially another chapter, as mentioned above. It is well placed as an appendix since it deals mainly with computer graphics. The idea behind virtual space lies in the algorithms created by a programmer-designer. If we were asked to put together all the geometries known up to the present day, we would find ourselves in serious trouble. Without pursuing the solution of this conundrum, this is exactly what computer graphics has done. Computer graphics makes it possible to manipulate all kind of projections (orthogonal, perspective, etc.), and to construct any desired view of an image in 2D or 3D. Instead of the artist's hand, it is now an algorithm that easily controls the modeling process of a simple wireframe image until it is finally transformed into a hyperrealistic image. The known geometries, such as Euclidean, perspective, trigonometry, analytic, descriptive, fractal, affine, polyhedra, all seem to become as one through the invisible language of computers. My conclusion after reading this chapter is that, nevertheless, behind the new visual world created by the computer something is missing; only drawings done by hand can enhance our creativity while computers appear not to improve it. Computers are very powerful tools, but our brains are still more powerful, so leave the creative job to our brains and let computers do the rest.

I highly recommend reading *Forma y Representación*. Since the aim of the book is to give a general notion of all the subjects discussed, many topics regarding geometry and perspective are dealt with, with no pretense of going into each one of them exhaustively. The illustrations themselves become almost like a second book, suggesting themes of further investigation to the reader.

About the reviewer

Tomás García-Salgado received his professional degree (1968), Master's degree and Ph.D. (1981-1987) in architecture. He is a formal researcher in the Faculty of Architecture of the UNAM (México), and holds the distinction as National Researcher, at level III. Since the late 1960s, he has devoted his time to research in perspective geometry, his main achievement being the theory of Modular Perspective. He also has several works of art, architecture, and urban design. More information regarding his work is available at http://perspectivegeometry.com.

NEXUS NETWORK JOURNAL — Architecture and Mathematics

Copyright

Submission of a manuscript implies: that the work described has not been published before (except in form of an abstract or as part of a published lecture, review or thesis); that it is not under consideration for publication elsewhere; that its publication has been approved by all co-authors, if any, as well as – tacitly or explicitly – by the responsible authorities at the institution where the work was carried out.

The author warrants that his/her contribution is original and that he/she has full power to make this grant. The author signs for and accepts responsibility for releasing this material on behalf of any and all co-authors. Transfer of copyright to Springer (respective to owner if other than Springer) becomes effective if and when the article is accepted for publication. After submission of the Copyright Transfer Statement signed by the corresponding author, changes of authorship or in the order of the authors listed will not be accepted by Springer.

The copyright covers the exclusive right and license (for U.S. government employees: to the extent transferable) to reproduce, publish, distribute and archive the article in all forms and media of expression now known or developed in the future, including reprints, translations, photographic reproductions, microform, electronic form (offline, online) or any other reproductions of similar nature.

All articles published in this journal are protected by copyright, which covers the exclusive rights to reproduce and distribute the article (e.g., as offprints), as well as all translation rights. No material published in this journal may be reproduced photographically or stored on microfilm, in electronic data bases, video disks, etc., without first obtaining written permission from the publisher (respective the copyright owner if other than Springer). The use of general descriptive names, trade names, trademarks, etc., in this publication, even if not specifically identified, does not imply that these names are not protected by the relevant laws and regulations.

An author may self-archive an author-created version of his/her article on his/her own website. He/she may also deposit this version on his/her institution's and funder's (funder's designated) repository at the funder's request or as a result of a legal obligation, including his/her final version, provided it is not made publicly available until after 12 months of official publication. He/she may not use the publisher's PDF version which is posted on www.springerlink.com for the purpose of self-archiving or deposit. Furthermore, the author may only post his/her version provided acknowledgement is given to the original source of publication and a link is inserted to the published article on Springer's website. The link must be accompanied by the following text: "The original publication is available at www.springerlink.com".

The author is requested to use the appropriate DOI for the article (go to the Linking Options in the article, then to OpenURL and use the link with the DOI). Articles disseminated via ww.springerlink.com are indexed, abstracted and referenced by many abstracting and information services, bibliographic networks, subscription agencies, library networks, and consortia.

While the advice and information in this journal is believed to be true and accurate at the date of its publication, neither the authors, the editors, nor the publisher can accept any legal responsibility for any errors or omissions that may be made. The publisher makes no warranty, express or implied, with respect to the material contained herein.

Special regulations for photocopies in the USA: Photocopies may be made for personal or in-house use beyond the limitations stipulated under Section107 or 108 of U.S. Copyright Law, provided a fee is paid. All fees should be paid to the Copyright Clearance Center, Inc.,222 Rosewood Drive, Danvers, MA 01923, USA, Tel.:+1-978-7508400, Fax:+1-978-6468600, http://www.copyright.com, stating the ISSN of the journal, the volume, and the first and last page numbers of each article copied. The copyright owner's consent does not include copying for general distribution, promotion, new works, or resale. In these cases, specific written permission must first be obtained from the publisher.

The *Canada Institute for Scientific and Technical Information (CISTI)* provides a comprehensive, world-wide document delivery service for all Springer journals. For more information, or to place an order for a copyrightcleared Springer document, please contact Client Assistant, Document Delivery, CISTI, Ottawa K1A 0S2, Canada (Tel.: +1-613- 9939251; Fax: +1-613-9528243; e-mail: cisti.docdel@nrc.ca).

Subscription information

ISSN print edition 1590-5896
ISSN electronic edition 1522-4600

Subscription rates

For information on subscription rates please contact:
Springer Customer Service Center GmbH
The Americas (North, South, Central America and the Caribbean)
journals-ny@springer.com
Outside the Americas: subscriptions@springer.com

Orders and inquiries

The Americas (North, South, Central America and the Caribbean)
Springer Journal Fulfillment
P.O. Box 2485, Secaucus, NJ 07096-2485, USA
Tel.: 800-SPRINGER (777-4643), Tel.: +1-201-348-4033
(outside US and Canada), Fax: +1-201-348-4505
e-mail: journals-ny@springer.com

Outside the Americas

via a bookseller or
Springer Customer Service Center GmbH
Haberstrasse 7, 69126 Heidelberg, Germany
Tel.: +49-6221-345-4304, Fax: +49-6221-345-4229
e-mail: subscriptions@springer.com
Business hours: Monday to Friday
8 a.m. to 8 p.m. local time and on German public holidays

Cancellations must be received by September 30 to take effect at the end of the same year.

Changes of address: Allow six weeks for all changes to become effective. All communications should include both old and new addresses (with postal codes) and should be accompanied by a mailing label from a recent issue.

According to § 4 Sect. 3 of the German Postal Services Data Protection Regulations, if a subscriber's address changes, the German Post Office can inform the publisher of the new address even if the subscriber has not submitted a formal application for mail to be forwarded. Subscribers not in agreement with this procedure may send a written complaint to Customer Service Journals, within 14 days of publication of this issue.

Back volumes: Prices are available on request.

Microform editions are available from: ProQuest. Further information available at: http://www.il.proquest.com/umi/

Electronic edition

An electronic edition of this journal is available at springerlink.com

Advertising

Ms Raina Chandler
Springer, Tiergartenstraße 17
69121 Heidelberg, Germany
Tel.: +49-62 21-4 87 8443
Fax: +49-62 21-4 87 68443
springer.com/wikom
e-mail: raina.chandler@springer.com

Instructions for authors

Instructions for authors can now be found on the journal's website: www.birkhauser.ch/NNJ

Production

Springer, Petra Meyer-vom Hagen
Journal Production, Postfach 105280,
69042 Heidelberg, Germany
Fax: +49-6221-487 68239
e-mail: petra.meyervomhagen@springer.com
Typesetter: Scientific Publishing Services (Pvt.) Limited, Chennai, India
Printers: Krips, Meppel, The Netherlands
Printed on acid-free paper
Springer is a part of
Springer Science+Business Media
springer.com
Ownership and Copyright
© Kim Williams Books 2010